PENGUIN BOOKS

FILTERS AGAINST FOLLY

Garrett Hardin is Professor Emeritus of Human Ecology at the University of California, Santa Barbara. He is the author of numerous books, including *Nature and Man's Fate, Exploring New Ethics for Survival, Stalking the Wild Taboo,* and *Naked Emperors.*

Garrett Hardin

FILTERS AGAINST FOLLY

How to Survive Despite Economists, Ecologists, and the Merely Eloquent

PENGUIN BOOKS

PENGUIN BOOKS
Published by the Penguin Group
Viking Penguin Inc., 40 West 23rd Street,
New York, New York 10010, U.S.A.
Penguin Books Ltd, 27 Wrights Lane,
London W8 5TZ England
Penguin Books Australia Ltd, Ringwood, Victoria, Australia
Penguin Books Canada Limited, 2801 John Street,
Markham, Ontario, Canada L3R 1B4
Penguin Books (N.Z.) Ltd, 182–190 Wairau Road,
Auckland 10, New Zealand

Penguin Books Ltd. Registered Offices:
Harmondsworth, Middlesex, England

First published in the United States of America by
Viking Penguin Inc. 1985
Published in Penguin Books 1986
Reprinted 1987

3 5 7 9 10 8 6 4

Grateful acknowledgment is made to *Liveright Publishing Corporation* and *Granada Publishing Ltd.* for permission to reprint the lines from "voices to voices, lip to lip," from *IS 5,* poems by E.E. Cummings. Copyright 1926 by Horace Liveright. Copyright renewed 1954 by E.E. Cummings.

LIBRARY OF CONGRESS CATALOGING IN PUBLICATION DATA
Hardin, Garrett James, 1915–
Filters against folly.
Reprint. Originally published: New York, N.Y. :
Viking, 1985.
Includes index.
1. Human ecology. 2. Environmental policy.
3. Economic policy. I. Title.
GF50.H36 1986 304.2 85-29028
ISBN 0 14 00.7729 4

Printed in the United States of America
Set in Baskerville

PREFACE

Among the greater glories of our civilization is the impersonal support society sometimes offers to those who would modify its institutions. Philosophers and scientists of an earlier day (Leibnitz, for instance) had to spend much time and effort tacking to the fickle winds of patrons or hoped-for patrons. At the least, such behavior is undignified. More serious is the fact that constant kowtowing blunts the ability to observe objectively.

In most of the twentieth century the major financial support of inquisitive minds has come from universities and foundations. This has given scholars a degree of independence unmatched in any other century. (Whether this will continue now that federal subsidies are taking over the role of private institutions is a serious question left for time to answer.) I have been fortunate enough to live and work in the freer part of the century.

Most ecology, like most science, is routine and unthreatening, but at the deepest level ecology deserves the label given it by its wise elder statesman, Paul B. Sears: "the subversive science." Ecology's most profound insights call for far-reaching modifications of long-standing social arrangements. It takes intellectual independence to achieve and voice such insights, as well as financial support to make the intellectual work possible. The University of California at Santa Barbara has given me most of that support, with significant augmentation in recent years by the Ecumenical Fellowship of Los Angeles. I am most grateful to both institutions, and to their decision makers.

This book had its origin in a single lecture prepared for a

Nobel celebration at the Gustavus Adolphus College in St. Peter, Minnesota. The spirited discussion that followed led me to believe that the thesis deserved to be treated on a more extensive scale.

Criticism is an indispensable need of scholarship. Since I am primarily interested in recasting the knowledge of specialists into lay language, my greatest source of useful criticism has been audiences in universities, colleges, and civic organizations. In addition to the lectures that were individually arranged, I have greatly benefited from lecture tours set up by the honor societies of Phi Beta Kappa and Sigma Xi. My debt to the numerous and anonymous disputants thus encountered is immense.

In the writing of this book, the greatest critical help has come from my wife, Jane, and my editor, Alan D. Williams. The printed result must be my thanks to them: I trust it is adequate. I have had great satisfaction in composing this work; I hope others experience at least half as much stimulation in weighing the arguments contained herein against their particular experiences.

CONTENTS

PART II. THE LARGER VIEW 139

PART I

THE THREEFOLD WAY

CHAPTER 1

Not Vices But Properties

Passionate action, being close to the marrow of our animal nature, must be older than dispassionate inquiry, but the record shows that the ability to be dispassionate and objective is nearly as old as writing itself. In addition, the experience of anthropologists with the people of "primitive" cultures—"our contemporary ancestors," as they have been called—suggests that objectivity may be older than literacy. The suggestion can hardly be verified.

Be that as it may, the tender flower of objectivity is easily crushed by what is taken to be the necessity of the moment. Has the percentage of the population capable of dispassionate inquiry followed any trend over the course of time? It is hard to imagine the evidence that would establish this conclusion. More important: can education (in the broadest sense) increase objectivity in the general population? It is the faith of most professional educators that it can; but this is faith, not established fact.

Every new enthusiasm brings some backward steps in the march toward objectivity. Our generation has experienced an "ecological revolution" that has at times attracted more enthusiasm than understanding. Upon discovering ecological malfunctions in the organization of society, some enthusiasts have not hesitated to attribute them to wickedness on the part of the people in power. Consider, for example, this request addressed to the editor of *Sierra* (the bulletin of the Sierra Club) in May 1984:

> Could you please publish a concise, unemotional, factual summation of all the atrocities, and proposed atrocities committed

> by this administration? I don't think many people really know how vicious Reagan has been.

Little good is to be expected of an inquiry encumbered at the outset with such emotion.

The writer's language stands in sharp contrast to that used more than three centuries earlier by the philosopher Spinoza in his unfinished *Tractatus Politicus.* In laying out the ground rules for what we now call "political science," Spinoza wrote:

> That I might investigate the subject matter of this science with the same freedom of spirit we generally use in mathematics, I have labored carefully not to mock, lament, or execrate human actions, but to understand them; and to this end I have looked upon passions such as love, hatred, anger, envy, ambition, pity, and other perturbations of the mind, not in the light of vices of human nature, but as properties just as pertinent to it as are heat, cold, storm, thunder, and the like to the nature of the atmosphere.

Not vices, but properties—it is to the latter that we must look as we try to make sense of the ecological revolution. As we explore the differences of the contending parties, we can be quite sure that not all the sense is on one side, with all the nonsense on the other. "Vice" and "virtue" are seldom useful concepts. We want to know how social systems work, what the technical facts are, and how unexamined assumptions can produce results we don't want.

In our highly technological society we cannot do without experts. We accept this fact of life, but not without anxiety. There is much truth in the definition of the specialist as someone who "knows more and more about less and less." But there is another side to the coin of expertise. A really great idea in science often has its birth as apparently no more than a particular answer to a narrow question; it is only later that it turns out that the ramifications of the answer reach out into the most surprising corners. What begins as knowledge about very little turns out to be wisdom about a great deal.

So it was with the development of the theory of probability. It all began in the seventeenth century, when one of the minor French nobility asked the philosopher-scientist Blaise Pascal to devise a fair way to divide the stakes in an interrupted gambling game. Pascal consulted his lawyer-mathematician friend Pierre de Fermat, and the two of them quickly laid the foundation of probability theory. Out of a trivial question about gambling came profound insights that later bore splendid fruit in physics and biology, in the verification of the causes of disease, the calculation of fair insurance premiums, and the achievement of quality control in manufacturing processes. And much more.

I believe the pattern of this history is being repeated with ecology. Agitation about rather narrow environmental issues—pollution, principally—evoked a deep examination of distribution theory. Every human activity produces both things that we want—"goods"—and things we don't want—"bads." How should society distribute these goods and bads? Who should pay, and who should benefit? What are the remote consequences of this distribution system or that? And how remotely should we inquire?

Economists and philosophers have dealt with distribution problems for centuries, but it is now apparent that they did not do their homework well enough. A new corps of experts from the sciences is moving into the field of distribution, but they are not being welcomed. They are often perceived as interlopers, pirates, squatters. So they are—but in socially important questions, academic specialties cannot be granted exclusive property rights.

The organic growth of probability theory into questions of insurance, quality control, and the design of experiments is now being matched by the natural extension of ecological insights into such questions as the design of political systems, the rational limits of foreign aid, and the control of the suicidal tendencies of some of our so-called virtues.

At this point in the evolution of civilization the service of experts is indispensable. But each new kind of expert introduces new possibilities of error. It is unfortunately true that experts are generally better at seeing their particular kinds of trees than the forest of all life. Thoughtful laymen can become very good at seeing the forest, particularly if they lose their timidity about challenging the experts. When I speak of laymen ("laypersons," if you prefer) I am talking about everyone, because the expert in one field is a layman in all others. In the universal role of laymen we all have to learn to filter the essential meaning out of the too verbose, too aggressively technical statements of the experts. Fortunately this is not as difficult a task as some experts would have us believe.

What follows is one man's attempt to show that there is more wisdom among the laity than is generally conceded, and that there are some rather simple methods of checking on the validity of the statements of experts.

CHAPTER 2

The Expert as Enemy

I well remember the summer my wife and I went to Alaska to attend the annual meeting of the Pacific division of the American Association for the Advancement of Science. The year was 1969. Because bids were being submitted at this time for oil leases on Prudhoe Bay, many of the sessions of the AAAS were devoted to the possible economic, environmental, and social consequences of drilling oil at the rim of the Arctic Ocean and transporting it across the width of the state. A pipeline would have to be built, so the feasibility of this was also under discussion.

One of the delightful features of science meetings in this American outpost is the wide variety of people who attend. The usual assortment of professors and industry representatives is enriched with politicians, Indians, Aleuts, and Eskimos—groups seldom seen at science meetings in the "Lower 48." Discussion in Alaska frequently jumps out of the well-worn ruts.

One of the sessions was concerned with the expected effects of the proposed trans-Alaska pipeline. A lean, intense engineer made a sales pitch for the conglomerate scheduled to build the pipeline. His slide presentation was slick and convincing. The audience was bombarded with turgid tables and computer-generated graphs. The speaker gave the impression that every conceivable contingency had been dealt with.

Nevertheless, one layman was not convinced. This man, who had the rugged look of a modern sourdough, made his points very deliberately. "Let me see if I've got this straight. You say this 800-mile-long pipe, four feet in diameter and completely above ground for most of the distance, will be filled with oil which will be at 190 degrees when it enters the pipe at the northern end, right? And that the oil will cool continuously as it flows south, but it will still be 130 degrees, and fluid, when it reaches Valdez. Right?"

"Yes, that's right. Our computer model shows that the oil at Valdez will still be hot enough to flow during the coldest winter Alaska has ever had."

"What if there is some sort of traffic delay at Valdez and the oil has to stay in the pipe for a while, cooling all that time?"

"No problem. If there is a shortage of ships the oil can be held in the pipe for two weeks without solidifying. And remember, we have a tank farm at the south end to act as a buffer for the oil stocks."

"Yes. But suppose all the storage tanks are full, and then there is a tankers' strike that lasts *three* weeks. What happens then?"

The engineer's reply was cool and prompt: "That's an interesting question."

There is this to be said for the expert's response: it was honest. A politician in the same situation would no doubt have replied with some such words as these: "I'm so glad you asked that question. It gives me an opportunity to bring up another matter of the utmost importance to patriotic Americans, namely . . ." and his well-lubricated tongue would have generated a rhetorical smoke screen that would have eloquently submerged the original question. In blessed contrast, the engineer did not hide the fact that he had no solution to the difficulty.

Indeed, there is no general solution. In trying to picture the real situation, we must set aside the neat computer curves of the engineering analysis and picture a huge pipe that may, some frigid winter's day, be filled with 800 miles of something not too different from the tar of a blacktop road. No pumps in the world can move such a mass through so long a pipe. Even if we had the necessary pumps, the pipe could not withstand the pressure (particularly not the Alaska pipe, in the making of which—as we learned after it had been put in place—there had been cheating on the quality controls).

So what should a pipeline supervisor do when he gets the word, one unhappy winter's day, that the oil flow is going to be shut down at Valdez for an indeterminate period?

If the supervisor did nothing, and the shutdown lasted longer than two weeks, the pipe would become plugged with a viscous and immovable mass. Summer would not cure the ill: the Arctic summer is too short, and the angle of the sun too low to furnish the needed calories. There would be no practical way, ever, to heat the 800-mile pipe enough to make its contents fluid again. To avoid having to build a completely new pipeline, the supervisor would have to open valves somewhere and release the oil before it congealed.

But release it where? The pipe holds more than nine million barrels of oil. Can you picture the devastation such a large quantity of oil would cause as it flowed out over the tundra? Alternatively, picture what nine million barrels of oil would do poured out over the ocean fishing grounds around Valdez.

Such are the pictures that were missing from the engineer's computer curves. The expert's response, "That's an interesting question," gave no hint of the reality.

Well, the pipeline was built. In more than a decade of operation no serious accident has occurred. But some day we may learn the answer to the citizen's question.

Tunnel Vision vs. the Panoramic View

Belief in progress—technological progress—became something of a religion throughout the Western world in the nineteenth century. Not until the latter part of the twentieth century was there any substantial questioning of this religion. Openly expressed doubts of technology often evoke the emotional reaction accorded the public doubting of the established church in earlier centuries.

When commitment to a technology like the Alaska pipeline is opposed, it is usually assumed that the opponents are attacking the technology itself. This is seldom true. It may be granted that the technology is excellent, but technology does not exist apart from human beings. It is human beings who calculate dimensions, turn blueprints into structures, monitor the construction and operate the machinery, and report to the general public. *Fallible* human beings are involved at every critical point in the system. Our questions should take account of this reality.

Should the computer be trusted? *Who chose the program? Who logged in the data?*

Do the equations give an answer we can trust? *Who chose the parameters? Why were these parameters selected and not others?*

Are the machines reliable? *Who put them together? Were the workers emotionally upset by marital squabbles? Were they fighting with the shop steward? Was the union trying to undercut management?*

Why should we doubt the inspection log? *Quis custodiet ipsos custodes?* This is the old Roman way of asking, *Who inspects the work of the inspector? And the work of the inspector of inspectors? And the inspector of inspectors of inspectors?* . . . Thus is the yearning for reliability thwarted by an infinite logical regress.

In the real world no regress is truly infinite. At some point the drive to reliability must be brought to a full stop. At that point we face naked human nature in all its mystery and complexity. What appeared at first sight to have been a doubting of technology turns out to be a healthy appreciation of the complexity and unpredictability of human nature.

You don't have to be an expert to have a healthy respect for human nature. In fact, laymen often are better at evaluating reliability than are the experts. The expert who looks for "the light at the end of the tunnel" can all too easily end up with tunnel-vision. The layman whose attention is less narrowly focused may be better at taking in the whole panorama.

In Search of Filters

How are we laymen to survive in a world increasingly dominated by experts? It is all too easy for an expert minority to override the lay majority in the making of decisons.

First, we must get over the embarrassment of being called laymen. The root "lay" comes from the Latin *lacum,* meaning "lake" or "pool." Laymen, then, are all the people who belong to an undifferentiated pool of people, as opposed to the few who are specialized in some particular competence. Since everybody is a layman in most areas let us accept the term proudly and see what measures we can take to extract truth from the depositions of experts.

The greatest folly is to accept expert statements uncritically. At the very least, we should always seek another opinion. Moreover, to the extent that time allows, we may become a little bit expert ourselves; but we don't have time enough to go far in this direction. (If we went all the way we would become just one more expert, thus moving ourselves to the other side of the debate and leaving the rest of the lay public with the same old problem.)

We need lay defenses against expertise. Fortunately there are such. The most important defense measure is to make oneself sensitive to the biases introduced by the assumptions and methods of experts. The layman in the Alaska story showed such a sensitivity: he realized that the engineer's exclusive focus on quantitative data that could be plugged into his computer program led him to ignore less predictable elements, such as strikes and sabotage, which cannot be convincingly quantified and entered into engineering equations. The omission undercut the credibility of the conclusions.

Layman or expert, each of us is forever trying to put together a true picture of that indefinable something we call "reality." The world is too complex for our minds to encompass the whole. We must filter the data and arguments that come in to us. Filtration clarifies: that's good. But the clarification is achieved by filtering out—setting aside—part of reality. Different filters clarify—alter—the total picture in different ways. We need to know the characteristics of our filters, their merits and their defects. Only by understanding these will we be able to protect ourselves against the assumptions (conscious and unconscious), the biases and the prejudices, of our experts.

As for the poorest of the experts, his first failing is that he uses just one kind of filter. His second failing is his unawareness of his first failing.

Nowhere is the need to understand our reality filters greater than in the vexed area of the environment. The word "environment" is as hard to define as the word "reality." In the present context "environment" refers to two sorts of relationships. First, there are the relations of human beings to the nonhuman world, a world that makes no allowance for human desires or human needs. The discipline of ecology studies these relationships. Second, there is what might be called the internal environment in which human beings are the environment for other human beings. This world is the object of study of economics, political science, and ethics. The combined study of both kinds of relations is the province of human ecology.

A generation has passed since the environment became a controversial issue. The most important differences of opinion have developed between economists and ecologists. I am inclined to echo Shakespeare in saying, "A plague on both their houses," though I live in one of them.

Many of the controversies can be resolved by looking closely at the reality filters used by the contending parties. Human beings are the instruments of scientific progress. Technical results are often difficult to understand, but the peculiarities of the instruments—fallible human beings—are both fascinating and understandable by all. In what follows, concern with human nature takes precedence over the details of technology.

An expositor who wants to convince should confess his vested interest at the outset. Let me do so. I am a card-carrying ecologist; as such, my view of reality is no doubt sometimes distorted by unconscious biases. I can hardly claim otherwise. Nor can I tell others where I am wrong: if I knew that, I would mend my ways *now.* I can, however, point out some of the errors of my ecological colleagues (though they won't thank me for it). There's enough blame to go around for all the specialists to share in the plenty.

In earlier days human beings often mistreated the world and "got away with it." People—*some* people—survived each environmental disaster, though the world as a whole was impoverished. The Tigris-Euphrates river system now supports only a fraction of the population that once flourished there. Bad managers "got away with" criminal mismanagement, because in those days there was a lot of world and not much humanity (in both of two senses).

Now there is a surplus of demanding human flesh and daily less of the unspoiled environment needed for the nourishment of that human mass. In the category of "nourishment," I include both physical nourishment and something more subtle than the physical. If I weren't allergic to the word, I might call this kind of nourishment "spiritual." Whatever the best adjective, I have in mind the inexpressible delight that comes from being acutely aware that one is breathing clear mountain air, drinking from uncontaminated and unchlorinated streams, swimming at isolated beaches, or contemplating great natural beauties that we hope to save for our grandchildren. Awareness is an essential part of experience.

Despite the thrust of the last paragraph, the discussions that follow will not often be directly concerned with such standard environmental topics as pesticides, contaminated water, and smoggy air. The principal concern will be with the very human impulses that make it so difficult for us to reach agreement on the management of environmental amenities.

Why is it so difficult to save something for our children?

What, if anything, can be done about poverty? How much compassion can we afford? Why does the noble dream of "One World" so easily lead to disaster? Can the rules of private philanthropy be converted into public policy without generating tragedy? Can the many contradictory meanings of "conservative" be reconciled? Is there a scientifically defensible form of conservatism?

The past three decades of controversy over environmental matters have shown that ancient philosophical questions can be profitably reexamined in the light of these scientific facts. In the end, the principal benefit of the ecological analysis that has come out of the environmental movement may be that it makes us more aware of the boundaries set by human nature. In the end, we can realistically hope to use our reality filters with greater facility.

CHAPTER 3

Three Filters of Reality

Controversies have two sources: a struggle for power, and an impulse to resolve puzzles. Some "realists" would reduce every dispute to a pure clash of animal ambitions. I think this is too cynical. Despite all our ego-centeredness, in the end we are at least somewhat amenable to logical arguments. Allowed sufficient time—often, admittedly, an unconscionable amount of time—ego may yield to reason. It is when animal ambition refuses to yield that the worst follies of humankind are perpetrated.

On the grand scale, folly produces tragedy. In the fictional realm there was Oedipus, who closed his ears to Jocasta's pleas to give up the investigation into his past. In history so recent that most Americans still remember it, there were the tragedies of Watergate and Vietnam. In her book *The March of Folly,* historian Barbara W. Tuchman speaks of the invariable "wooden-headedness" of those headed for tragedy. As description the term is apt, but it is no explanation.

It would be naïve to expect a unitary explanation of so widespread a human characteristic as wooden-headedness. Bull-headedness (to use another term) can in fact be adaptive up to a point: in such case we may call it "steadfastness." It is the excess that needs guarding against.

Explaining maladaptive behavior is not enough: we also want to avoid it. We want prophylactics against folly. How do we get trapped in foolish actions to begin with? It is my contention that many of our mistakes can be both understood and avoided if we are acutely aware of three intellectual filters available to us in our dealings with reality. The three-filter analysis presented here grew out of an earlier two-culture analysis.

Snow's Two Cultures

What is the proper role of science in the making of society's decisions? Should science be merely an errand-boy to human ambition, or should its insights help shape our desires? More succinctly, should science be "on tap or on top"? In 1956 an English novelist and physicist-turned-administrator, C. P. Snow, sharpened this question by distinguishing two cultures in Academia. His remark-in-passing was followed three years later by the publication of a brief and beautifully written essay, "The Two Cultures." Despite much criticism, the distinction made by Snow has become part of the conventional wisdom.

To those outside universities, Academia may seem a unity. It appears far otherwise to those inside. Not only are scholars separated from each other by high departmental walls but the disciplines are also aggregated into two regrettably antagonistic groups. These, Snow called "cultures," though anthropologists might have preferred the term "subcultures." Whatever the label, one of Snow's groups consists of mathematics and the natural sciences, the other of the humanities, i.e., history, literature, philosophy, and the like. (The disjunction fits the English Academia better than the American, which has schools of business, physical education, industrial arts, et cetera, areas not considered proper academic subjects in the British isles.) Ideas are the basic concern of both cultures. Nevertheless an eavesdropper, listening to daily conversations in both cultures, hears more discussion of *words* in the humanities and more of *things* in the sciences.

Students of the humanities ruled the universities long before the study of things became respectable for anyone but lowly artisans. Real science traces back at least to ancient Greece, but it was only rather recently that the many individual sciences were gathered together under the single umbrella "science." The Latin precursor, *scientia,* meant simply knowledge, any sort of knowledge. In the English language it was not until the eighteenth century that the present meaning of "science" became established. As late as the mid-nineteenth century, science was quite often referred to as "natural philosophy." (Since philosophy literally means the love of knowledge, *natural* philosophers were those who pursued knowledge about nature, about the things of the world.) English-speaking people had to wait until 1834 for the word "scientist" to be coined.

During the first half of the nineteenth century most English scientists had to find their occupational niche outside the universities. Science, still called "natural philosophy," was fighting for a place in the curricula of Oxford and Cambridge. It was nearly 1860 before an Oxbridge student could major in a science. The dons, who determined which students were given fellowships, were not favorably impressed by students who excelled in science, referring (on one occasion) to a top score in exams in natural philosophy as "a dirty little first in Nat Phys, whatever that is."

The attitude that regards work with things—work carried out with the hands—as dirty is one that is reborn time after time in societies prosperous enough to support a class that uses its skill with words to belittle those who make things work. Each such birth sets in train the potential suicide of society. Scholars' deprecation of working with the hands is still a serious roadblock to scientific advancement in Latin America, where entrenched upper classes honor poets more than scientists. The style of the humanistic culture—no manual labor—has infected the scientific culture in Latin America. Latino scientists commonly assign all the routine work of the laboratory (of which there is much) to lowly untrained helpers. Thus do they deprive themselves of the subtle but real benefits that only "hands on" experience can give. National leaders in Latin America wonder why their science is so unproductive. To scientists in other countries the answer is obvious.

Two Meanings of "Culture"

The reciprocal isolation of the two cultures in the academic world is apparent even in the history of the word "culture." The word grew out of the older agricultural term "cultivation." In the nineteenth century two new meanings developed. In 1871 E. B. Tylor gave "culture" the anthropological meaning around which C. P. Snow later constructed his argument, namely (to quote the *American Heritage Dictionary* of 1969) "the totality of socially transmitted behavior patterns, arts, beliefs, institutions, and all other products of human work and thought characteristic of a community or population."

In 1876 (five years after Tylor), the English poet, essayist, and educator Matthew Arnold defined culture as "the acquainting ourselves with the best that has been known and said in the world." In a number of essays he made it clear that "the best that has been known and said" refers to contributions from the humanistic side only.

The two definitions overlap but they are significantly different. Arnold's culture is a value-laden term; Tylor's culture is merely descriptive. *Every* identifiable group of people possesses an anthropological culture, no matter how wretched its daily life. In contrast, Arnold's type of culture may be entirely lacking in a society that refuses to be concerned with what connoisseurs might call "the best that has been known and said."

Words play a more important role in Arnold's culture than in Tylor's, for the "acquainting" and "known and said" in Arnold's definition clearly imply the use of words. On the other hand, the anthropological definition merely presumes that ideas are transmitted from one generation to another. This is quite a different matter, for the transmission of ideas, practices, assumptions and ideals may be carried out using numerous techniques of "nonverbal language" as well as by writing and speaking.

We are all inescapably immersed in anthropological culture. In contrast, the humanist's culture is a Sunday thing, an embellishment of life that may or may not be present. Part of the controversy that followed the publication of Snow's essay arose because it did not occur to some of the Arnoldian critics that the word "culture" might have another meaning. As evidence, note that the authoritative 12-volume *Oxford English Dictionary,* published in 1933, gives space to Arnold's kind of culture but not to Tylor's. That the anthropological meaning of culture could still be lacking in a dictionary 62 years after Tylor, gives some measure of the isolation of Snow's Two Cultures. Not until the publication in 1972 of the A–G Supplement to the OED—101 years after Tylor, and 16 years after Snow—was the word "culture" treated adequately from an anthropological point of view.

What Is an "Intellectual"?

Clarity is praiseworthy, but it can intensify antagonisms. Certainly the clarity of Snow's essay did. The irritation created in some quarters by his basic thesis was exacerbated by his suggestion that we should reexamine the customary meaning of the word "intellectual." *Wordsmiths*—journalists and those academics whose lives are built around words—long ago succeeded in getting this eulogistic word accepted as a label for people who often are no more than merely eloquent. By this definition, many great mathematicians and scientists would have to be classified as nonintellectuals, for a significant proportion of them (not all) are extemely poor at explaining their ideas in words. Snow attacked the traditional definition of "intellectual" by pointing out that most so-called intellectuals from the Arnoldian culture had no knowledge or appreciation of the meaning and awe-inspiring implications of the Second Law of Thermodynamics, though this giant among concepts was already a century old.

(Happily, in the two decades since Snow's essay, a number

of popular treatments of entropy and thermodynamics have diminished the ignorance. Perhaps Snow's essay was partly responsible for this welcome and long overdue advance.)

Snow alienated many in the humanistic culture by claiming that the professional's knowledge of the essentials of the other culture was far greater among scientists than it was among literary and artistic people. He was in an unusually favorable position to make such a claim, since he moved in both circles, being a scientist by vocation and a successful novelist by avocation.

Focus on Filters

Snow's approach can be legitimately criticized on several grounds. To begin with, his skill with words may have enabled him to create a "self-fulfilling prophecy," thus furthering the very polarization of the intellectual community that he deplored. Polarization and objectivity don't mix well; the search for truth becomes submerged by the implicit and demanding question "Are you with us or ag'in us?"

More important, Snow's categories are far from exhaustive. Academic people are inclined to forget that most bright people are neither physicists nor poets. If "intellectual" refers to those who possess and employ a high degree of intelligence, then the inclusive class of true intellectuals must take in some businessmen, clerks, laborers, artists, homemakers—in fact, people from all occupations. An exact accounting might well show that the majority of true intellectuals actually live outside the so-called "intellectual community."

What we need most is a categorization not of people, but of the methods whereby people express and test statements. We need to be acutely aware of the virtues and shortcomings of the tools of the mind. In the light of that awareness, many public controversies can be resolved.

Though some metaphysicians demur, scientists assume that there is a reality "out there" to which scientific statements refer. Our intellectual tools are filters for reducing reality to a manageable simplicity. To assert that we have ever completely captured reality in an equation or a string of words would be extremely arrogant. Each filter captures only part of reality. Most expertise is single-filter expertise. We come closest to the truth when we compensate for the bias of one filter by using others (which have different biases).

How many intellectual filters men and women now employ, or may invent in the future, is probably beyond knowing. In my opinion, most of the major controversies of our time can be well understood if we view them as the result of too much reliance on any single one of the following three reality filters:

the literate filter
the numerate filter
the ecolate filter

By focusing on these three filters, rather than on Snow's two cultures, we have a better chance of keeping our attention directed toward substantive issues rather than toward personalities.

Distinguishing the Filters

For centuries literacy has been regarded as the prime distinguishing mark of the educated person. The ability to read and write enabled the individual to draw on the wisdom (and the foolishness) of human beings distant from him in space and time. The intellectual life of the literate person was thus more extended than that of people who were unable to decipher those curious marks on papyrus, clay tablets, or paper.

With the coming of radio and television the relative advantage of truly literate people diminished considerably. Like it or not, many people who today are no more than nominally literate have fairly accurate knowledge of what is happening all over the world, and even some understanding of what happened in the past. Many effective politicians and captains of industry don't read a book from one year's end to another. The documents they read do not at all correspond to what the Arnoldian means by "literature." Many a man of action receives most of his information through the ear.

For this reason I will henceforth use an extended definition of "literacy" to stand for *skill in either written or spoken language*. A person adept in either will here be considered *literate*. By virtue of modern technology the advantages and shortcomings of the two forms of literacy are today much the same. A literate person is competent in the fine scale analysis of words and their meanings, and adept at finding the best word to express his ideas. Consciously or not, literate analysis begins with this question: "What are the words?" i.e., what are the most appropriate words?

Scientists, focusing primarily on things and processes, often pay little attention to words. (If they pay too little attention they can easily muck things up.) Scientists are apt to pay more attention to quantities and rates than they are to rhetorical niceties. (A rate is a quantity of change per unit time.) Numbers, in the broad sense, dominate scientific thinking.

Shortly after the end of World War II someone coined a term to epitomize this orientation, this sort of ability: "numeracy." Most scientists are strongly *numerate*. They seek to understand the world by passing reality through a numerate filter. The implicit question of the strongly numerate person is this: "What are the numbers?" i.e., what are the exact quantities, the proportions, and the rates?

The relative emphasis on numeracy and literacy roughly matches the distinction between Snow's two cultures. In 1970, looking deeper into the way that people analyze the world, it occurred to me that ecologists in particular use another filter, the filter of "ecolacy," to give it a name consonant with the other two.

Consider for instance the theory behind the evolution of resistance to antibacterial drugs by bacteria. At the literate level there is no reason to expect this evolution: by definition, an "antibacterial" kills bacteria, so that should be that. At the numerate level, the initial studies may show that all but 0.0001 percent of the bacteria are killed by the first application, so it looks like two applications should kill all but $(0.0001)^2$ = 0.00000001 percent, and surely that should be enough? But if two applications are not enough, how about three?

The simple numerate analysis does not suggest the whole truth, which is that continued application of the antibacterial substance often results in the appearance of a new form of bacteria that is immune to the drug (as the original form was not). Something else is going on.

That something else is natural selection operating against a background of random genetic variation. The one bacterium in ten thousand that survives the first application is not a random sample of the total bacterial population: it is the rare variant that has a natural resistance to the drug. It survives and breeds more of its own kind. Continued exposure to the drug causes the bacterial population to evolve toward drug resistance. Differential death accomplishes the unwanted miracle. The pessimistic generalization that comes out of such experiences is this: "Every pesticide selects for its own failure." This is an essentially *ecolate* doctrine: neither literacy nor numeracy would lead one to predict it.

The ecolate result may involve more than the target species. Insecticides often have a delayed effect of causing an unwanted increase of the targeted species because the existence and importance of other species have been overlooked. For instance, spraying for aphids may result in a delayed outburst of these pests, because the insecticide also kills ladybird beetles, insects that feed on aphids. The natural world is organized into a web of life more complex than we know. We have only a limited ability to predict what will happen *in time* as the result of any intervention, however well meant, in the natural order of things. Caution and humility are the hallmarks of the ecolate attitude toward the world.

Ecolacy, the Time-Bound Insight

Time and its consequences are essential concerns of the ecolate filter. The key question of ecolate analysis is this: "And then what?" That is, what further changes occur when the treatment or experience is repeated time after time?

Time and the web of life thwart many attempts to remake the natural world nearer to human desires. We thought we could wipe out malaria in Africa and Southeast Asia by spraying mosquitoes with DDT. In fact, after an initial apparent success, we found we were selecting for DDT-resistant mosquitoes. Malaria is now increasing in those areas.

The tunnel vision of enthusiasts seldom encompasses either time or the reticulated web of organisms and causes. The ecolate critic, with his irritating question "And then what?", is not welcomed by enthusiasts, whether they be profit-minded promoters or altruistic reformers. The commercial promoter wants to make a fortune selling something, e.g., pesticides. The reformer wants to diminish human suffering through single-minded attention to only one of the factors involved: for example, through increasing food production in a poor country in which the environment is being ruinously exploited by an overfertile population. Such a reform increases the number of people, which in turn has one of two effects: either the amount of food per capita ultimately decreases, or the population becomes increasingly dependent on donations of food from the outside.

A cautious approach to innovation is the mark of a true conservative. With respect to the things he understands best, an ecologist is clearly a conservative (whatever may be his attitude in other areas). But is this not the behavior of genuine experts in every field? Anti-conservatism (under whatever name) is more often associated with ignorance than with knowledge. Prudence, an essentially ecolate virtue, is not popular with enthusiasts.

In summary, the three filters operate through these particular questions:

Literacy: *What are the words?*
Numeracy: *What are the numbers?*
Ecolacy: *And then what?*

No one filter by itself is adequate for understanding the world and predicting the consequences of our actions. We must learn to use all three. This means we had first better understand the strengths and weaknesses of each.

CHAPTER 4

Sins of the Literate

Eight years after entering political life, the Victorian novelist Edward Bulwer-Lytton wrote (in his play *Richelieu*) one of English literature's best-remembered half-remembered couplets:

> Beneath the rule of men entirely great,
> The pen is mightier than the sword.

When a man who is competent in both politics and literature gives pride of place to the former, we should take his judgment seriously. But how often does one hear the entire quotation? Practically never: the people most given to quoting are wordsmiths, and the first line does not properly massage their egos.

Speech has long been correctly praised as the sovereign sign of humanity, as that which, in the idiom of Bulwer-Lytton's day, "separates man from the brutes." Thanks to recent researches we now know that other animals also have languages—of a sort. Although the new knowledge has diminished our species-conceit somewhat, we still find the subtlety and power of human speech awe-inspiring. However, since these virtues have been extolled quite enough in the past by those whose livelihood is tied to language, it is time to offset the old bias with a new. From the encounter of conflicting biases something closer to the truth should emerge.

In discoursing on the seriousness of literary biases with a friend from the world of letters, I once offered in evidence a remark that Robert Lowell is recorded to have made. The occasion was one on which the significance of W. H. Auden's work was under discussion. "After all," said Lowell, citing Auden's moving poem "September 1, 1939," "if not for Auden we wouldn't have known about the Second World War."

Taking the words at their face value I deduced that Lowell's bias was such that he felt that nothing in this world was real until it had been suitably treated in literature. Not at all, said my friend: "Don't you sense the irony?"

I didn't, and there's the trouble. Language no doubt began as straight description of the world, but it soon grew to encompass the indirect approaches of irony, sarcasm, and contrafactual conditionals. To understand what is meant, one often has to be able to hear two languages: language in the ordinary sense, and the unspoken language that tells you how to "hear" the spoken. The second "language" is often called "meta-language." An important accomplishment of the literary artist is his ability to make the reader "hear" unspoken meta-language. It hardly needs saying that a competent artist with words does not preface an ironical comment with the statement, "I will now speak ironically." It would spoil the effect.

Nothing like irony is to be found in the literature of science. The loss in piquancy is compensated by a gain in precision.

Many literary people are content with a world that is rather remote from things. Some of them even have an antipathy to things. The essayist Norman MacLean once tried his hand at writing some Western fiction stories, without much commercial success. As one unsympathetic editor explained on a rejection slip: "These stories have trees in them." The comment is not significantly different from the classic remark of a movie actor turned politician: "When you've seen one redwood tree, you've seen them all." We must conclude that neither editor nor politician had much love for the things of nature.

No doubt most people would agree that the principal function of language is to promote communication between people, perhaps adding that it serves also to further that interior communication we call thinking. All well and good, but we do not truly understand the social significance of language until we explicitly express this greater truth: *Beyond communication, language has two functions: to promote thought, and to prevent it.*

Worshippers of language ordinarily mention only the first function, but occasionally they let the cat out of the bag. "The true use of speech," said Oliver Goldsmith, "is not so much to express our wants as to conceal them." The thought-prevention function of language is most obvious in statements made by promoters bent on thwarting the law. Consider, for example, the public relations problems of "massage parlors." A noneuphemistic label would provoke trouble with the law, so one ingenious owner of a parlor in Manhattan's Times Square named his facility the "Fellowship for Human Happiness," incorporating it under the New York State Religious Corporation Law. By defining it as a religious institution the owner escaped certain taxes.

The verbal camouflage, had it been too successful, could have interfered with attracting the trade the owner wanted, so corrective leaflets were handed out on the street extolling the services offered by the "church." The male customer was invited to "be captivated by your own personal, glamorous tranquillity angel in the seclusion of our temple." The disguise escaped legal challenge for more than two years.

"Keeping in Touch," the Excuse of Logorrhea

"Speech," said Thomas Mann, "is civilization itself. The word, even the most contradictory word, preserves contact. It is silence that isolates." There's something in this, for the human species is, after all, a social species. But with words, as with all goods, there can be too much of a good thing. Too many words can get in the way of understanding.

One of the great promoters of the idea of evolution in the nineteenth century was the British philosopher Herbert

Spencer. Historically he was perhaps as important as Charles Darwin in selling the idea to the educated public, yet today his name is seldom mentioned in biology texts. Why not?

Biologists find no mystery in this neglect. Spencer simply smothered his ideas with words; with Spencer, facile writing was often a substitute for hard thinking. Consider his definition of evolution:

> Evolution is an integration of matter and concomitant dissipation of motion; during which the matter passes from an indefinite, incoherent homogeneity to a definite, coherent heterogeneity; and during which the retained motion undergoes a parallel transformation.

Within a short time an English mathematician, T. P. Kirkman, gave an apt paraphrase of this definition: "Evolution is a change from a nohowish, untalkaboutable all-alikeness, to a somethingelsification and sticktogetherations." Incensed, the philosopher devoted three pages to an erudite refutation of this parody. From this, and numerous other instances, we are justified in concluding that Spencer suffered from a severe case of "logorrhea"—verbal diarrhea.

Whether or not the verbiage is excessive, esteemed writers have often muddied the waters of discussion by confusing what should be recognized as distinct categories of existence. Those who claim sovereignty over the discussion of "spiritual" matters usually delight in condemning the "materialism" of others. Yet ironically, they often try to endow the immaterial entities they postulate with the properties of material things. Does the fact that the word "substantive" is a synonym for "noun" lead the incautious to assign the status of *substance* to abstractions that are originally, by definition, insubstantial? If not, what else can explain the following aberrations?

> "Faith is the substance of things hoped for, the evidence of things not seen."—*Hebrews* 11:1

> "Reason is substance, as well as infinite power, its own infinite

> material underlying all the natural and spiritual life; as also the infinite form which sets the material in motion. Reason is the substance from which all things derive their being."—Hegel
>
> "Spirit is the only Eternal Absolute Substance. Nature is an outward and visible sign of this inward-underlying Energy or Being. Its phenomena are naught else than the *objectified modes* of the Eternal I Am."—Thomas Howard MacQuery

The confounding of insubstantial entities like "faith," "reason," and "spirit" with true substances has a charitable explanation in the dating of the quotations from a time when the scientific distinction between material and immaterial entities had not been clearly made. But that is no excuse for prolonging the confusion.

Infinity as a Thought-Stopper

Coming back to the point that one of the functions of language is to prevent thought, we should note how immensely useful the word "infinity" is in preventing people from thinking about quantities and their implications. "Infinity," together with its derivatives "infinite" and "infinitesimal," has a certain utility in mathematics, but its technical limitations are subtle and best left for mathematicians to deal with. When incorporated into everyday writing "infinity" usually leads to trouble. This is so even when the word itself is avoided, provided the *idea* of infinity is implied. Consider the following instance from the U.S. Bureau of Soils, in 1909:

> The soil is the one indestructible, immutable asset that the nation possesses. It is the one resource that cannot be exhausted; that cannot be used up.

And from a recent opponent of abortion:

> Moral law spells out the sanctity of life in very specific terms, attributing *infinite value* to every innocent life. Infinity is indivisible. Any fraction of infinity remains *equally* infinite.

And there is this from the Secretary of Defense, testifying before a Senate committee in 1967, in response to the question, "How long can the U.S. afford the gigantic financial cost of the major ground war in southeast Asia before our economy breaks down?" To this the Secretary responded in just three words: "I think forever."

Though the word "infinite" is used in only one of the quotations, the idea of infinity is central to all. To see the unity in the arguments we need to employ a simple technique made explicit by the physicist Percy Bridgman early in this century. (The method was used implicitly, though more or less unconsciously, long before by critical thinkers in science and outside.) The method is called *operationism,* or "operationalism." Faced with conflicting views, the critical analyst asks, "What operations are implied by these statements?" Once the operations are made clear, difficulties usually evaporate.

In recasting the statements just given in operational terms, no attempt will be made to determine the actual desirability of soil conservation, abortion, or the Vietnam war. We are concerned here only with the legitimacy of the idea of "infinity" in its various verbal disguises. Had the authors been completely candid, their statements would not have been significantly different from the ones given below. (It is hardly likely, however, that they would agree to these, or to any other, operational paraphrases.)

> Agricultural bureaucrat: "It is ridiculous to suppose that the soil could ever be exhausted: *I refuse to talk about it.*"
>
> Abortion opponent: "I am not interested in assertions that abortion may sometimes be the lesser of two evils: in fact, *I refuse to talk about it.*"
>
> Secretary of Defense: "Don't bother to try to figure out the

domestic consequences of continuing to pursue the war in Vietnam: *I refuse to talk about it.*"

In the real world of material resources, infinity, as the philosopher Whitehead observed, "has no properties. All value is the gift of finitude." To the man possessed of a trillion trillion dollars (just to keep the figure nominally finite), a single dollar would have no verifiable personal value, and there would be no psychologically perceptible difference between one dollar and ten . . . or a hundred . . . or a thousand. Value is the gift of finitude.

If the soil providentially renewed itself no matter how we mistreated it . . . if all children were wanted, no matter when born, and if all parents had an infinite amount of loving care to dispense . . . and if the means of waging war and preserving domestic peace had no true limits—then, indeed, it would be no more than sensible to put an end to further discussion by introducing the word "infinity" or one of its equivalents.

But the resources of the world—*our* world—are finite. On this point ecologists and other scientists are agreed. Mainstream economists also agree that the world in which economic decisions take place is finite, repeatedly quoting the axiom "There is no such thing as a free lunch." Thus have economists aligned themselves with the great tradition of the natural sciences, which are, at all times, committed to the discipline of "conservation-thinking," that is, to closed systems of reckoning. Matter and energy are jointly "conserved": there is never, in human experience, any creation or destruction of matter/energy.

In contrast to the orientation of the majority, a few economists on the fringe have in recent years spun out some rather skillful rhetoric, the thrust of which is that there is a free lunch out there somewhere. Or, as one of the cleverest of them has put it, the natural resources available to human beings "are not meaningfully finite." On the face of it, this statement is meaningless; it is, however, immensely appealing to those to whom finitude is threatening. In so "hard-nosed" a science as economics it is hard to believe that this minority view will long endure. In the meantime it gums up public discussions.

The Protean Nature of Discussion-Stoppers

During the troubled times of the 1960s young men and women, trying to direct the attention of their elders toward the serious problem of the Vietnam war (as well as toward the less serious problem of the governance of university campuses), invented a new discussion-stopper: *non-negotiable.* A long list of demands would begin or end with the statement "These demands are non-negotiable." Such a term makes civilized discourse impossible. Its use bespeaks a cowardly desire to escape the intellectual discipline of civilized discussion. Those who use such terms as *non-negotiable, self-evident, must,* and *imperative* try to deny others the right of responding.

He who introduces the word "infinity" or any of its derivatives ("forever" or "never," for instance) is also trying to escape discussion. Unfortunately he does not honestly admit the operational meaning of the high-flown language used to close off discussion. "Non-negotiable" is a dated term, no longer in common use, but the word "infinity" endures forever (if you will pardon the expression).

Like old man Proteus of Greek mythology, the wish to escape debate disguises itself under a multitude of verbal forms: infinity, non-negotiable, never, forever, irresistible, immovable, indubitable, and the recent variant "not meaningfully finite." All these words have the effect of moving discussion out of the numerate realm, where it belongs, and into a wasteland of pure literacy, where counting and measuring are repudiated. It would be reasonable for readers and auditors to demand a frank acknowledgment of the operational intent of the users of such sonorous verbiage; but the demand would surely be met only with a multiplication of more innumerate words, for the following reason.

We human beings are egotistical animals; each of us wants to win the argument. Each of us wants to escape opposition. But if *I* openly reveal my egotistical desire *to* you, I may evoke a similar assertion *from* you. . . . Stalemate!

However, if I can make you think I am talking about the properties of things outside my egotistical self—about the infinite quantity of some nonpersonal thing—I may succeed in keeping you from perceiving the egotistical base on which my argument is built. . . . He who thinks language is *only* a form of communication is badly deceived.

Above all else, *language is action:* it serves the demands of the ego whether it is used to promote, or to prevent, thought.

The Problem of Poetry

Yet it must be admitted that words of the infinity family can cast a lovely patina over literary passages. Consider for instance this moving claim by Shakespeare's King Richard II:

> Not all the water in the rough rude sea
> Can wash the balm off from an anointed king.

What a lovely way to put into words the idea of the unlimited and divine right of kings! Richard asserts that the endurance of this right is infinite. (Translation: "I don't want to talk about it!") There was a time when this sovereign discussion-stopper was acceptable.

No more. Most people now understand the ruinous consequences of political power that is forever unexamined, a power never weighed in the balance. The charisma of leadership may still exert its intoxicating effects, but it is unsupported by any mystical infinity. We still anoint our leaders with the balm of office, but the balm is easily washed off.

Does this mean that we must forgo the pleasure of listening

to Shakespeare? Is it hypocritical to enjoy poetry that is rich in thought-stoppers like "infinity," while at the same time we assert our commitment to a rational, scientific approach to life's problems?

Not at all. The poet Coleridge (who, incidentally, had a keen appreciation of the science of his day) has given a formula for justifying the poetic impulse as well as the poetic product. Whenever he put together a poem, he said, he endeavored to "procure for these shadows of imagination that willing suspension of disbelief for the moment, which constitutes poetic faith." In this passage there is no denial of rationality and no permanent commitment to the irrational. *Only for the moment* is there a willing suspension of disbelief—and it is a suspension, not a rejection of the critical impulse.

Nearer our own time A. E. Housman has disagreed with those who claim too great an empire for poetry: "I cannot satisfy myself that there are any such things as poetical ideas. . . . Poetry is not the thing said but a way of saying it."

Poetry is least dangerous when the typographical arrangement of words reveals the author's poetic intent. It is most dangerous when the argument is cast in the form of prose, in sentences heavily infected with unacknowledged poetic claims of non-negotiability. In our time the claims of recognized poets are no longer a serious threat to rational thought. The gravest threats to rationality now come from those who employ the rhetorical weapons of poetry from behind an ambush of prose. Popularizers of ecology and advocates of the environment are not the least of the offenders. Consider, for instance, this statement by John Muir:

> The universe would be incomplete without man; but it would also be incomplete without the smallest transmicroscopic creature that dwells beyond our conceitful eyes and knowledge.

One can grant that this is "true by definition," that is, by the definition of "completeness"—but so what? We need to augment Muir's thesis with some figures before we can see the box it puts us in.

There are something like ten million species of plants and animals in the world, the greatest concentration of them being in the tropics, where our knowledge is least certain. "Developers" are now in the process of destroying one habitat after another in the "undeveloped" world. With each such destruction, thousands of species are eliminated. It has been authoritatively estimated that one-third of the world's species will be extinguished between now and the year 2000, if present trends continue.

The estimates are very rough, but that doesn't matter. A little arithmetic shows that anything like this rate of destruction eliminates about two dozen species every hour. On the average, it took the evolutionary process about a million years to produce each existing species from its progenitor species. The disparity of these two rates is shocking.

Most of the species are small, most are unknown to most people, and many will not be known even by the specialists before they disappear. A society that cherishes variety should be concerned about what is happening. And some of the species are no doubt important for the welfare of humankind, even from the most narrowly practical point of view.

But what are we to make of Muir's claim that the world is "incomplete" when even a single species is eliminated? Every two or three minutes, if our figures are even approximately correct, it becomes more incomplete. In a year's time, does our universe become 200,000 times as incomplete as it does every two or three minutes? Is the concept of "completeness" the best abstraction to serve as a handle for this problem?

Most biologists, especially ecologists, think there is a problem. We think that the world would be richer, more beautiful, and better off if we succeed in preserving most of the species. (Not necessarily every last one, however: it now appears that we may have eliminated the virus of smallpox around the year 1980. If time shows that this elimination was complete and final, who among us regrets that the world is now more "incomplete" by reason of the elimination of this "transmicroscopic creature"?)

Extinctions caused by human activities present a snarl of ethical, esthetic, practical, and political problems. Conflicting goals have to be balanced one against another. There is no royal road to rationality. Whatever means we are tempted to use, we must be wary of the poetic approach. Rhetoric like Muir's may give one a wonderful "oceanic feeling" (to use Freud's term), but this feeling is more likely to prevent than to facilitate advances in understanding. It is when ecological rhetoric is most beautiful that we must be most on our guard.

CHAPTER 5

The Numerate Filter

The word "numeracy" has only recently made its way into English dictionaries. Finding I am dissatisfied with published definitions, I offer my own:

> **numeracy,** *n.* **1.** The art of putting numbers to things, that is, assigning amounts to variables in order that practical decisions may be reached. **2.** That aspect of education (beyond mere literacy) which takes account of quantitative aspects of reality.

No one connected with business—and most adult Americans are connected with business in one way or another—needs to be convinced of the central importance of numerate thinking. Just as certainly, those who have their roots in science know how often a *qualitatively* new view grows naturally but surprisingly out of *quantitative* investigations. Galileo, in the seventeenth century, set the tone of modern science when he said, "The Book of Nature is written in mathematical characters." Whitehead, in our century, agreed: "Through and through the world is infected with quantity. To talk sense, is to talk in quantities."

The love of numeracy and mathematics has esthetic roots. The poet Edna St. Vincent Millay, though no scientist, captured some of the rapture in her sonnet, "Euclid alone has looked on beauty bare." This aspect of scientific beauty is largely absent from popularizations of science. A typical television presentation glories in the complexity of the apparatuses it shows, missing the point that quite frequently the principal purpose of an expensive machine may be to produce only a single reliable figure, needed to settle a complicated controversy.

A British astrophysicist, the late Sir Arthur Eddington, was fond of pointing out that much scientific investigation could be successfully carried out by a Cyclops with tunnel vision, capable of taking only one meter reading at a time. From an aggregation of such minimalist Cyclopean reports the vast and beautiful mental pictures of the structure and workings of the universe can be assembled.

Achieving much with little: this esthetic ideal is as widespread in science as it is in any of the activities traditionally recognized as arts. The successful result is called "elegance."

Euclid was not the only one who looked on beauty bare; to a greater or lesser extent all scientists are similarly privileged. In numerous popular lectures in the first part of the twentieth century Eddington tried to convey the esthetic excitement of scientific progress—from pointer readings to cosmology, as it were. One wonders if it was hearing one of Eddington's lectures that led e. e. cummings to protest:

> (While you and i have lips and voices which
> are for kissing and to sing with
> who cares if some oneeyed son of a bitch
> invents an instrument to measure Spring with?

It would appear that some people have trouble seeing the beautiful in the bare.

Mathematical Machismo

The widespread antipathy to quantitative science is, in part, merited. The more arrogant of the practitioners of numeracy have harmed the cause. For example, the Victorian physicist Lord Kelvin smugly asserted:

> When you can measure what you are speaking about and express it in numbers, you know something about it, and when you cannot measure it in numbers, your knowledge is of a meagre and unsatisfactory kind.

Biologists take a malicious satisfaction in noting that Lord Kelvin ridiculed the evolutionists of his day for believing that the world is many hundreds of millions of years old; Kelvin claimed that his mathematical analyses indicated an age of no more than a few million years. On firmer ground scientists today estimate an age of between 5,000 and 20,000 million years. Why was Kelvin so spectacularly wrong?

Assuming no careless error has been made, the conclusion of a mathematical analysis is only as good as its premises. Kelvin calculated the age of the earth by figuring out how long it would take a newly formed planet of such a size, at such and such an initial temperature and at such a distance from the sun, to cool to its present temperature. What Kelvin did not know at the time he made this calculation (because no one knew it) was that radioactive elements in the depths of the earth produce heat. During all the millions of years the earth was losing heat through radiation into space, it was also gaining heat from the decomposition of radioactive elements. When this became understood in the twentieth century, physicists corroborated the rough calculations of the biologists of the preceding century.

(Those who take pleasure in seeing the mighty fall might like to know that Kelvin just as confidently asserted that man would never fly in a craft that was heavier than air. He also predicted that any metal cooled nearly to the temperature of absolute zero would become an electric insulator. Truth contradicted the natural philosopher: as K. Onnes found in 1911, supercooled metals become superconductors. Ironically, the zero point on the absolute temperature scale is called "zero degrees Kelvin.")

Numeracy More Than Measurement

Just as "literacy" is used here to mean more than merely reading and writing, so also will "numeracy" be used to mean more than measuring and counting. Examination of the origins of the sciences shows that many major discoveries were made with very little measuring and counting. The attitude science requires of its practitioners is respect, bordering on reverence, for ratios, proportions, and rates of change. Rough and ready back-of-the-envelope calculations are often sufficient to reveal the outline of a new and important scientific discovery.

G. N. Lewis has given a description of scientific method that stands in sharp contrast to the restricted view of Lord Kelvin. Lewis, a great physical chemist and an even greater teacher of chemists in the twentieth century, declared: "I have no patience with attempts to identify science with measurement, which is but one of its tools, or with any definition which would exclude a Darwin, a Pasteur or a Kekule." The physiologist Walter B. Cannon extended Lewis's list with the names of Harvey, Virchow, Pavlov, and Sherrington. In truth, the essence of many of the major insights of science can be grasped with no more than a child's ability to measure, count, or calculate.

The numerate temperament is one that habitually looks for approximate dimensions, ratios, proportions, and rates of change in trying to grasp what is going on in the world. Given effective education—a rare commodity, of course—a numerate orientation is probably within the reach of most normal people.

The resources of literacy include too much rhetoric that is beautifully suited to the hiding of numbers and the need for numbers. Dichotomies are favored over quantities. It is so comforting to divide polluting substances sharply into the categories of "safe" and "unsafe." Mutually exclusive categories make the writing of regulatory laws simple, of course, but nature seldom draws a sharp line at which "unsafe" begins.

"Where is the dividing line between safe and unsafe speeds for an automobile?" . . . "What is a safe amount of radiation for a human being?" . . . "At what level should we start worrying about acid rain?" . . . Nature is silent. Nature does not tell us when "safe" slips over into "unsafe"; men and women, reasoning together, must legally *define* "unsafe."

Wherever the line is drawn in law, the decision is an arbitrary one. A speed of 55 miles per hour is not significantly and provably more dangerous than 54.473 mph; we merely agree to begin taking legal notice of vehicle speeds above that number. Thus it is with all numerical standards of safety. The final point of the decision is always arbitrary.

The wording of the previous paragraph brings up a subtle point of difference between the scientific culture and the merely literate. In science "arbitrary" is used to indicate a decision made in order to get on with the work; it carries no aura of disapproval. Usage in science is closely related to the legal meaning recorded in the dictionary: "Dependent on the discretion of an arbiter or other legally recognized authority; discretionary, not fixed." Arbitrariness is an adaptive response to the practical need for action.

Among some groups, particularly among the more radical of social activists, the word "arbitrary" can be counted upon to evoke a knee-jerk reflex of strong emotional disapproval. In these circles a different dictionary definition is assumed: "Derived from mere opinion; capricious; not limited by law; absolute, despotic." The conflict between Snow's "Two Cultures" often arises because, quite unconsciously, they are operating with two different dictionary definitions of the arbitrary.

Life is impossible without arbitrary decisions (however unconscious the decision maker may be of his dependence on the arbitrary). But so strongly do some people reject all risk in life that they make decisions that in the end make life riskier. The passage of the Delaney Amendment is a prime example of the passion for security gone berserk.

The Delaney Amendment

Five years before the book *Silent Spring* was published, Congress passed a scientifically indefensible law, the Delaney Amendment to the Pure Food and Drug Act. Concerned with the growing evidence that many otherwise useful substances can cause cancer, Congress decreed that henceforth, whenever a chemical at *any* concentration was found to cause cancer—in *any* fraction of *any* species of animal—that substance must be totally banned as an additive to human food. Such is the "zero tolerance" criterion written into the law.

The Delaney Amendment is a monument to innumerate thought. "Safe" and "unsafe" are literate distinctions; nature is numerate. Everything is dangerous at some level. Even molecular oxygen, essential to human life, becomes lethal as the concentration approaches 100 percent.

Completely banning any substance is likely to lead to legal, administrative, and economic problems sooner or later. Scientists are continually increasing the sensitivity of their measuring instruments. Sensitivity is ordinarily expressed as "1 part per X," where X is a large number. If a substance provably increases the incidence of cancer at a concentration of 1 part per 10,000, one should probably ban it at that concentration in food, and perhaps at 1 in 100,000. But what about 1 part per million? That used to be a common limit of the sensitivity of measuring methods, but now 1 part per billion is often achieved.

(Before going further I call attention to my use of figures—"1," "2," "3," et cetera—instead of the spelled-out words "one," "two," "three," et cetera, as decreed by style books of the literate culture. Words do not capture the mind as figures do. . . . This may seem a small matter but the history of science shows that symbolization is an important element in intellectual progress. The battle for numeracy begins at the desk of the copy editor.)

In theory, there is no final limit to sensitivity. What about 1 milligram per tank car? Or 1 milligram per terrestrial globe? With sufficient sensitivity in our measuring instruments we could be driven to conclude that nothing is safe to eat. But the alternative to eating is fasting, and that is lethal, too.

Why was the scientifically indefensible Delaney Amendment enacted into law? A psychoanalyst might suggest that the high average age of congressmen, particularly in the 1950s, produced an immoderate fear of cancer, which led these otherwise intelligent men to abandon rationality. This is not the sort of hypothesis that can be easily proved, but it is plausible. Irrationality needs some explanation.

The curse of the Delaney Amendment is still with us. The Delaney Amendment should either be repealed or (better) restated on a quantitative basis. To achieve their maximum utility, safety standards must be numerate.

Stupidities like the Delaney Amendment would be committed less often if we felt in our bones the truth that Paracelsus asserted more than four centuries ago: "All things are poisons, for there is nothing without poisonous qualities." Those who agree with Paracelsus are not surprised to learn that our most valuable antibiotics, analgesics, anesthetics, and tranquilizers have bad effects on some people at some concentrations under some conditions of use. All effective medicines are poisons to some extent. Beneficent poisons, if you wish, but poisons nonetheless: they should be used with quantitative discretion, with suspicion, and with vigilance.

Quantities matter. Numbers matter. Duration of time matters.

Our attitude toward pollution also needs to be passed through a numerate filter. Had all environmentalists been imbued with the numerate spirit of science, there would have been fewer controversies during the past quarter of a century. Unfortunately, many of the enthusiasts who moved into the environmental movement after *Silent Spring* brought with them an almost wholly literate orientation. They detested pollution and craved purity. Absolute purity. They wanted to enforce zero tolerance on all environmental pollutants, not just on carcinogens. Such innumerate enthusiasts would, if they could, universalize the Delaney disorder.

With friends like these the environment needs no enemies.

CHAPTER 6

The Pursuit of Reliability

The controversies that followed the publication of Rachel Carson's *Silent Spring* in 1962 more often than not sprang from a misunderstanding of the true nature of the author's concerns. Her criticisms were not so much directed at technology itself as they were at the reliability of the human beings who install and operate the technology. Carson tried unsuccessfully to forestall misunderstandings with this passage at the end of her first chapter:

> It is *not* my contention that chemical insecticides must never be used. I do contend that we have put poisonous and biologically potent chemicals indiscriminately into the hands of persons largely or wholly ignorant of their potentials for harm. [Italics added.]

That human beings are fallible has been known since the beginning of time, but modern technology adds new urgency to the recognition. It is essential that there be a sense of numeracy among those who employ technology. Directions for using pesticides are always numerate: X amount of chemical is to be dissolved in Y amount of solvent; the mixture to be applied at the rate of Z amount per acre per unit of time.

Technology is used by human beings whose reliability can be expressed as a fraction. *Fraction,* because convention dictates that perfect reliability be indicated by the numeral 1, while experience shows that all human reliabilities are equal to or greater than 0 but less than 1. "Only Allah is perfect," as the Moslems say, so the reliability of human beings—or human-made machines—is never as great as 1. Rachel Carson implied that it is not Allah who distributes chemical pesticides over the landscape. This is hardly a radical assertion, but industry spokesmen expressed shock and incredulity.

Human reliability can never be raised to the divine number 1. Technologies with the greatest promise for good sometimes bring with them equally great potential for harm when improperly used. Nuclear power and poisonous pesticides are examples. Such a coupling of promise and risk may seem hardly fair, but, as cannot be said too often, "Life isn't fair."

Sellers of chemicals and machinery who want policy to be set solely on the basis of the reliability of their products miss the point. The reliability that matters is not the simple reliability of one component of a system, but the final reliability of the total control system. The total reliability is the product of two factors:

$$(\text{Technology reliability}) \times (\text{Human reliability}) = (\text{System reliability})$$

Or, abbreviating:

$$\text{T.R.} \times \text{H.R.} = \text{S.R.}$$

To make the significance of this equation more obvious, we replace the abstract entities with particular fractions, thus passing from the literate form to the numerate. Arbitrarily, let us assume that T.R. equals 0.9, while H.R. equals 0.6; this means that the probability that technology will not fail us, over a certain period of time, is 90 percent, while the probability that the human beings involved will not let us down, over the same period of time, is 60 percent. Since both T.R. and H.R. are necessarily involved in the system, the probability that the total system will behave properly is the product of the two partial probabilities, or:

$$0.9 \times 0.6 = 0.54$$

Note that the reliability of the total system is less than that of the less reliable component.

Suppose we manage to increase the value of T.R. to 0.99? With such a 1 percent failure rate (per unit time) system reliability rises to 0.594—which is still less than H.R. alone.

More to the point, suppose we could persuade Allah to design and manufacture our technology, which means that T.R. would then equal 1.0. System reliability then becomes $1.0 \times 0.6 = 0.6$. The result may be seen as a special instance of the ancient wisdom that a chain is only as strong as its weakest link. Even without achieving a divine level of reliability, modern technology has been so greatly perfected that the weakest link in most technology/human systems is the human element.

We can look at this another way. Returning to the original values of T.R. = 0.9 and H.R. = 0.6, let us suppose that with equal effort and expense we could increase either component reliability by five percentage points. If, then, we want to conserve effort and expense, which improvement should we strive for? Let's compare the results:

(A) improve technology $0.95 \times 0.6 = 0.57$
(B) improve people $0.9 \times 0.65 = 0.585$

Obviously we should follow path (B), that is, devote our efforts to improving human reliability (since technical reliability is already so high). This conclusion is strengthened when we note the lack of realism in the assumption that equal improvements in T.R. and H.R. will follow from equal effort. The situation is analogous to the problem of carrying the football another yard when the ball is downed on the fifty-yard line versus a ball one yard from the goal: the closer to the goal, the more difficult it is to make a further yard. So it is also with high and medium degrees of reliability.

In most systems, the greatest theoretical potential for improving total system reliability lies in improving the reliability of human beings. But how are we to do that? Impatient reformers may be tempted to dream up totalitarian systems of education and control, but the ultimate control element in all totalitarian systems is a handful of human beings—who remain obdurately fallible. Even the most "reliable" of totalitarian systems—reliable, that is, over the short term of a generation or two—suffers from an inherent instability that eventually shatters it, with great suffering. If human reliability can be improved fundamentally (and it has not been proved that it can), the improvement must be made by something other than totalitarian means.

There is another way to improve system reliability: move some of the fallible elements from the human side of the system to the technical side, where we know better how to make improvements. This is what we do every time we devise a "foolproof" device. (There's a bit of wishful thinking in that time-honored term, but let it stand.)

An electrical system can be destroyed by a lightning bolt. We could station a man at a switch to break the circuit whenever there was a power surge—but the man might have just gone to the bathroom when the bolt struck. We bypass this difficulty by installing an automatic circuit breaker. Over the long run this technical device is far more reliable than human beings. This is but one example among tens of thousands that could be cited.

Obviously that's the way to go: redesign technology/human systems by moving components from the human side to the technology side. In fact, that's the way we have been going for a long time. But complete reliability still eludes us, for several reasons.

The replacement of fallible human beings by much more reliable technical devices is not a one-to-one replacement. The human animal is marvelously flexible and adaptable: he or she can do many things. No complex, skillful human action can be replaced by a single machine, so the replacement is many-to-one. A single human being is replaced by a complex assemblage of machine elements, each of which has to be manufactured by human beings, assembled by human beings, and tested and certified by human beings. Even when each link in a technical complex is much more reliable than direct human action, the final reliability may be no greater.

Suppose we want to replace a human element in a manufacturing process by a chain of machine elements. Suppose that the human element has a reliability of 0.6; and that each link in the technological chain has a reliability of 0.99 (1 percent failure rate), and none must fail. If the technological chain has two links in it, the reliability of the whole is:

$$0.99 \times 0.99 = 0.9801$$

If there are three links, the total reliability is the cube of 0.99, or 0.970299.

How many links must there be in the technological chain for the total technical reliability to fall to the human level (which we have supposed to be 0.6)?

Just 51. The decimal fraction 0.99 carried to the 51st power is 0.598956. The example is based on simplified assumptions, but the conclusion is general.

There is yet another door through which unreliability can sneak in. In no area of human endeavor have there been greater efforts to achieve 100 percent reliability than in the multi-billion-dollar space program. Yet even here, "only Allah is perfect." When the launch of space vehicle Gemini VI had to be aborted, the failure was traced to a dust cover that should have been removed before the launch. Investigation showed that a workman had signed a sheet that certified that he had removed the dust cover. Moreover two inspectors had further certified that the operation had been performed correctly. Would another ten inspectors have increased the reliability? It is doubtful.

Behind every complex machine there stands an invisible army of human beings. People make machines, people test them, people record the results of the tests, people examine the records, people certify to the examination of the records; and people reexamine the machines after use, record the results and certify the recordings, and so on. All inspection procedures suffer from an infinite regress. The paths to error are beyond numbering—and human beings are involved at almost every point along the way. Inspections are usually so easy to make that stupidity is seldom the reason for failed inspections. But it is boring to make the same inspections day after day, month after month. Certification is expensive. Laziness (on the part of inspectors) and greed (on the part of contractors) account for more failures than stupidity.

There is yet one more way in which systems can fail. On 9 November 1965, a massive electrical blackout covered most of New England; some of the communities tied into the great grid of the electric utilities were without power for several days. The trouble started with a single failure in a power station in upstate New York. The automatic transfer of power from one element to another in the system caused surges that touched off one circuit breaker after another—a true instance of the often-cited domino effect.

When the investigation of the blackout was completed the conclusion was simple: the massive failure had been a system failure. The logical characteristics of the electric grid system made failure inevitable at some unknown time. The very safety devices themselves had created a potential for failure in the system.

Of course, once the system deficiency was understood, the fault was corrected in similar systems throughout the country. That particular sort of failure should not occur again. But have the new system designs created new potentialities for system-generated failures? No one knows. No analyst is clever enough to be able to certify that the logical system of a power grid does not, under unthought-of conditions, contain the seeds of systemic suicide. The mathematical problem apparently exceeds the analytical abilities of the human mind. Here is yet another way in which we have to acknowledge and accept the fallibility of human beings. As we depend more and more on ever more complex control systems, we can hardly help being somewhat uneasy.

Out of Numeracy, Humanism

"Man is the measure of all things," said Protagoras in the fifth century B.C. Like many memorably succinct statements, the aphorism is ambiguous: depending on how the hearer interprets it, it seems either very wise or very foolish. In the largely technology-free world of Protagoras's time it no doubt seemed more true than it did some two thousand years later, after a rich development of technology.

Then, curiously, the further extension of technology made Protagoras's aphorism relevant once more. From now on, the more reliable our physical devices become, the more the measure of human reliability becomes the final measure of the reliability of the technology/human systems on which human welfare depends. By pushing technology to its furthest limits man has become more dependent on himself. A modern Protagoras would have to say, "Man is the *limit* of all human possibilities."

In reaching this conclusion we have used the resources of mathematics, but mathematics of the very simplest sort. Ratios have been employed—for that is what a reliability factor is (0.6 equals 6 out of 10, or an average of 6 successes out of 10 chances). Fractions have been multiplied together—a simple operation. Probability products have been compared. And that's about it. Numeracy of the simplest sort, but out of it comes the essentially humanistic conclusion that the only thing we can really count on in this uncertain world is human unreliability itself. We must learn to live with this inescapable fact.

Some of our most expensive technical experts have trouble adjusting to this humanistic conclusion. In the mid-1970s a $2,000,000 study of nuclear reactor safety, known as the "Rasmussen study," concluded that the probability of a significant reactor accident was less than the probability of a person's being hit by a meteorite. Unfortunately, in casting their analytical net the experts had failed to capture the human element. Human unreliability was, to the Rasmussen commission, no more than "an interesting question," *period.*

Then in March 1980 there occurred the accident at the Three-Mile Island reactor in Pennsylvania. The damage to human health was trivial, but the cost of repairing the plant ran into the billions of dollars. An accident of the type that occurred at Three-Mile Island was not foreseen in the massive Rasmussen study.

After investigating the reactor failure, a presidential commission concluded that ". . . except for human failures, the major accident would have been a minor incident." So the nation had spent $2,000,000 for a study that ignored the most important element of system reliability, namely the unreliability of the human animal.

There was a lot of numerate cleverness in the 3,300 pages of the Rasmussen report. But no amount of naked numeracy can make up for failing to understand that reliability is a function of a total system, and not the sole consequence of the reliabilities that are easiest to quantify. A different approach is needed to understand systems. This is the approach that engineers call "systems analysis" and biologists call "ecology."

And human beings are part of the total system.

CHAPTER 7

The Ecolate Filter

In the second half of the twentieth century ecology entered the consciousness of the general public through a door labeled "side effects." Remembering that one of the functions of language is to prevent thought, we easily recognize the purpose of the term "side effects": it is to discourage thinking about the total effects of a new medicine, a new pesticide, or a new public works project when some of the consequences prove embarrassing to the promoters.

But, to paraphrase Gertrude Stein, in the world beyond words, an effect is an effect is an effect. The adjective "side" is added to coerce thinking—to restrict questions to safe channels (safe for the promoter's enterprise).

How serious unforeseen side effects can be is apparent in the consequences of building the High Aswan Dam in Egypt. Completed in 1970, this dam on the Nile was designed to produce electricity and increase the supply of water available for year-round irrigation. The technical goals were clearly attainable, but long before construction started, at least one ecologist (Raymond F. Dasmann) pointed out some highly probable and highly undesirable consequences of controlling the flow of the Nile completely. Unfortunately, nobody was listening to ecologists in the 1950s.

What ensued turned out to be worse than feared. A brief review of the consequences—not to be called "side effects"!—is worthwhile. First, the perpetual wetting of irrigation channels made possible by the new dam favored the survival of snails and their parasites, thus augmenting the human toll of schistosomyasis, a seriously debilitating disease in tropical countries. In addition, throttling the flow of the Nile into the

Mediterranean stopped the deposition of silt in the delta, a process that had been making new farmland for centuries. Now normal erosion by the sea results in actual loss of previously deposited farmland. Unforeseen was the effect of decreased river flow on fisheries in the sea. The shrimp fisheries of the eastern Mediterranean, deprived of their yearly gift of nutrient-rich flood waters, have declined 97 percent. This has been hard on the fishermen. But worse is to come.

For five thousand years agriculture in the Nile Valley flourished because of the yearly deposition of a millimeter of rich silt during flood time. From now on that missing gift has to be compensated for by expensive chemical fertilizers. Moreover, without the yearly flushing of the soil by flood waters, salt will accumulate and eventually ruin the soil for farming. In probably less than a hundred years, five thousand years of successful agriculture will be brought to an end. Such a disaster deserves a more evocative name than "side effect."

During the same period of time the damming of the Volta River in West Africa has been followed by an enormous increase in the number of blackflies (simuliidae), together with their parasites, which cause "river blindness," a dreadful disease for both individuals and developing economies. In an attempt to control blackflies, DDT was used, but it killed the insects-enemies of blackflies as well, resulting in an even worse outbreak of river blindness.

Again: an epidemic of encephalitis occurred in Thailand after the Japanese invasion during World War II. Before that time water buffaloes took care of the heavy farm work, and disease-carrying mosquitoes pretty much confined their attention to these animals. When the Japanese substituted tractors for buffaloes, the mosquitoes turned their attention to pigs and human beings, becoming an important vector for an organism causing a disease that was soon christened "tractor-induced Japanese encephalitis."

Again: a similar sequence of events is believed to have taken place in Central Africa several centuries ago, when a more intensive form of agriculture was introduced. Greater disturbance of the soil led to the creation of more small puddles, bodies of water to which mosquito-eating fish had no access. As a result, tertian malaria increased greatly. The increased number of malarious natives acted as a powerful selective factor for inheritable malaria resistance in human beings. Among the genes turned up, providentially as it were, by the mutation process was one that causes human red blood cells to assume a sickle shape.

In the pure form ("homozygous state") the sickle-cell gene produces a disabling anemia. In the hybrid form ("heterozygous") it has little damaging effect, while conferring resistance to malaria. In spite of the human suffering caused by the homozygous state, the gene, on balance, does more good than harm to a population living in a malarious region without adequate medical knowledge of the cause and prevention of malaria. Natural selection favored the sickle-cell gene in Central Africa, and the gene became common over wide areas. In summary: an improved form of agriculture caused a great increase in anemia.

(Of course the advantages of the sickle-cell gene were lost when members of these populations were transplanted to non-malarious regions of the United States, or whenever public health programs eliminated malaria from African areas. Under such changed environmental conditions, selection today is eliminating the now wholly disadvantageous gene, but more slowly than people would like.)

The First Law of Ecology

The stories of these particular disasters can be matched by dozens of others, each of which could be labeled as Progress Gone Sour. Little is to be gained by lengthening the roster; what we seek are the general principles that guide ecology.

The essential orientation of ecology is all-inclusive, global: investigators try to take account of all factors at the same time. This is an ambitious undertaking: we should not be surprised if it encounters grave difficulties.

Experts, except for those operating under the banner of ecology, generally oppose the introduction of ecological questions into their problems. An expert is hired to carry out a limited analysis only. He naturally does not welcome another approach that undermines his conclusions. Engineers, who are very good at calculating the consequences of such purely physical factors as compressibility, ductility, and heat loss, don't like to be reminded of the possibility of labor strikes—as we saw in the story about the Alaska pipeline—or of carelessness, sabotage, revolutions, or those unnumbered forms of human behavior that are so unpredictable in detail but always so possible, Predictable or not, these possibilities are part of the total ecological system.

The ecologist attempts more, so he risks more. Any more closely specified predictions that he may make are more likely to be wrong than are those of a narrower specialist. But since his analysis encompasses a larger system, the consequences of ignoring ecological advice are likely to be graver.

There are those who maintain we have gotten along well enough without ecologists for many millennia: why complicate our lives by listening to them now? There are two answers to this question.

First, we have *not* gotten along nearly as well as most people think. Overwhelming evidence for this statement is presented in two massive volumes: *Man's Role in Changing the Face of the Earth* (1956), edited by William L. Thomas, Jr.; and *The Careless Technology* (1969), edited by M. Taghi Farvar and John P. Milton. When an area like that encompassed by the Tigris-Euphrates river system is forever ruined, the number of people living there is greatly reduced, their prosperity vanishes, and their influence on world opinion drops to zero. Who today in our part of the world knows the opinion of a single person living in the Euphrates valley? History is written by the winners: this natural bias is one of the great protectors of unjustified optimism. We who are momentarily on top pay little attention to the impoverished descendants of yesterday's losers.

The second reason for taking more seriously the predictions of the Cassandras of ecology is this: there are more people in the world now than ever before, though the world's physical resources are no greater than they were when *Homo sapiens* first appeared on the scene. We rub elbows more now. Like it or not, we are more often forced to take account of the consequences of our actions on others. As we spoil more of the environment, it becomes ever more important to consider the possibility of ruin. (Of course, technology effectively stretches the boundaries of the environment somewhat, but this gives us no reason to suppose that technology has no limits, even though we cannot specify them in advance.)

As guides to planning for an uncertain future, ecologists point to two statements made about a century ago. First there are some lines by the poet Francis Thompson:

> All things by immortal power
> Near and far
> Hiddenly
> To each other linkèd are,
> That thou canst not stir a flower
> Without troubling of a star.

The same thought was captured in a statement of the American naturalist John Muir: "When we try to pick out anything by itself we find it hitched to everything else in the universe." This is often shortened to: "Everything is connected to everything else." Whatever the wording, the idea is called the First Law of Ecology.

Ecology's big idea is global and exciting, but it is also (from a scientific point of view) dangerous. Neither Thompson's nor Muir's statement is numerate. Things without number are connected to each other, and measures of distance are presumed irrelevant. But the real world is not innumerate.

Take the matter of distance. Is it true that my nudging a

flower will affect the stars in the heavens? The nearest star, Alpha Centauri, is 25 million million miles away, and the only physical influences we know—e.g., gravity, magnetism, light—have their effects diminished by the *square* of the distance between objects. The poet's assertion is numerate nonsense.

Thompson could hardly have meant his poetic rhetoric to be taken that seriously as an assertion in the physical sciences. He was probably depending on the reader to read his lines with Coleridge's momentary "suspension of disbelief." When the reader does, no harm is done; but we must not forget the popularity of astrology, which makes such innumerate assertions in all seriousness. If the number of newspapers printing daily syndicated astrology columns is any indication, believers in astrology number in the millions in America. Of course, we don't know how many of them indulge in a Coleridgean sort of belief for only a few moments before returning to the world of numbers, quantities, and rationality. But unquestioning acceptance of poetic innumeracy seems, to scientists at any rate, like playing in the muck of irrationality.

It was considerations like these that led me, in 1963, to propose a modification of the First Law of Ecology:

We can never do merely one thing.

Perhaps the first thing that strikes the reader is that these seven words really do not assert very much. Quite true. But therein lies the merit that this variant enjoys over Muir's original. The vast majority of the world's connections are not *demonstrably* of importance. Seeking to uncover a multitude of undiscoverables can fritter away much time and effort. A person who embraces too "global" a version of the First Law may, as he becomes frightened by his impotence, settle for a comforting mysticism or fatalism, or the do-nothing attitude called "quietism."

Whatever adaptation we might make to Muir's thesis, the admirable commercial and civic energy that has characterized Western civilization in the past three centuries would be imperiled. I think it is the unconscious realization of the psychological consequences of taking Muir literally that has led many practical and productive people to reject ecology and environmentalism. They see the environmental movement as an emasculating force.

The wording "We can never do merely one thing" is less criticizable precisely because it asserts less. Rather than asserting an infinity of connections, it modestly implies that there is at least one unwanted consequence. It is much easier to watch out for a few things than for a multitude without number.

The new version has become widely accepted. Like the flag on a file folder, it keeps alive the consciousness of things that should not be forgotten. It is an echo of the famous warning of Oliver Cromwell: "I beseech you, in the bowels of Christ, think it possible you may be mistaken." The ecologists' warning is addressed to all those who are intoxicated with dreams of technology that may, in some small measure, prove to be mistaken.

Slippery Slopes and Camels' Noses

The Biblical "cloud no bigger than a man's hand" does indeed sometimes bring on a great storm. Back in the eighteenth century when the remote island of St. Kilda in the Outer Hebrides was still inhabited, the sight of an approaching sail was recognized by the residents as the forerunner of another epidemic of respiratory disease. The first cough of a disembarking sailor spelled trouble for the inhabitants.

Now for an example closer to our time and part of the world. The 61 definitive words of the Eighteenth Amendment to the American constitution—the Prohibition Amendment—created a new empire of crime, never completely liquidated by the annulment embodied in the Twenty-First Amendment.

Looking to the future, we may be sure that if we adopt a

constitutional amendment that gives legal recognition to the personhood of all human developmental stages from the fertilized egg on, the practical consequences will be fantastic. As of 1983, the number of induced abortions appears to have stabilized at 1.3 to 1.5 million per year in the United States. This is a measure of the demand. What will happen if this demand is thwarted? Given the record of the consequences of the Prohibition law it takes little imagination to write the scenario.

Clouds, coughs, and constitutional changes can have widespread effects. A small cause can have great effects.

But again, it may not. A cautious statement like the First Law of Ecology can alert us to danger, but it can also be used as an excuse for the most extreme sort of do-nothing conservatism. Three instances from nineteenth-century England will serve as examples: the outlawing of the use of "chimney boys" in the cleaning of chimneys, the mandating of the Plimsoll line (which indicates the capacity of ships), and the decision to manufacture both left and right boots for British soldiers in place of the "uniboots" that had been used for centuries.

Objectively, all three reforms were desirable. The first put an end to the early death of small boys from cancer induced by scraping their way through soot-lined chimneys. The second saved the lives of sailors by making it easy to deny departure permits to dangerously overloaded ships. The third made it possible for the British soldier to march farther with less discomfort by acknowledging what surely must have been obvious (even to generals) that a soldier's left foot is not identical with his right. Yet it took more than a generation to accomplish each of these reforms. Why?

First and foremost, because each reform threatened vested interests. Chimney sweeps saw their operating costs rising if they had to substitute technology for child labor. Ship owners figured that, over the long run, the extra revenue from overloaded bottoms would more than make up for the occasional loss of capsized ships. (In those days of nearly zero personal liability the lives of sailors cost nothing.) With respect to the soldiers' uniboots, there is no evidence that the military had any profit to gain from the non-sale of new boots. The more plausible explanation is that the generals had a vested interest in not thinking.

"Vested interest" was never offered as a defense of inaction, of course. The status quo was most often defended in the name of Free Enterprise (you could just hear the capital letters). It was repeatedly asserted that carrying out these reforms would be a first step in bringing about the downfall of the British Empire.

(Since England's empire has in fact dissolved by this time, perhaps the stand-patters were right after all? Cause-and-effect assertions about history are forever dubious because of the logical flaw of *post hoc ergo propter hoc* reasoning. Out of a multitude of contributing factors how can we be sure that a particular one is, or is not, determinative?)

The most astonishing story of all is the resistance offered to a liberalization of the marriage laws. Gilbert and Sullivan enthusiasts will recall that the Fairy Queen in *Iolanthe* decreed that young Strephon, going into Parliament, should "prick that annual blister/Marriage to deceased wife's sister." It is hard to believe now that the legality of marriage to a deceased wife's sister could ever have been under a cloud, but such was the case in England for centuries.

It took 125 years—from 1835 to 1960—for Parliament to remove completely the bars to such marriages. W. S. Gilbert helped the English people laugh about the ludicrous situation in 1882, but it is doubtful if he speeded up the reform.

Why the steadfast opposition to this change? Religious justification was found in *Leviticus* 18:16—"Thou shalt not uncover the nakedness of thy brother's wife"—which, by logical symmetry, was asserted to apply also to "thy wife's sister." Both were regarded as incest. This is biological nonsense, of course.

It can be argued that it is also religious nonsense. *Deuteronomy* 25:5 positively commands a man to marry the widow of his childless brother. Victorian moralists conveniently overlooked this passage. Selective attention was then, and is still, the rule among those who claim to found their ethical systems on Scripture.

Setting aside divine directives, why would anyone want to forbid a man to marry his dead wife's sister anyway? Since there is no biological relationship between the two, biological incest cannot be committed. The only plausible substantive danger cited was that of a man who finds his wife fading after bearing a child every year for several years, and who might understandably be attracted to her blooming younger sister. Lasciviously wanting to be free to start working on a more attractive sex object, he might be tempted to murder his worn-out wife. Possible. But how probable?

Matthew Arnold, a highly literate and certainly a wise man, probably put his finger on the real reason when he said that he feared that striking down the traditional prohibition would open the way to bigamy, polygamy, and what Gladstone described as "the more horrible forms of incest." In other words, read your Aesop: remember the tale of the Camel's Nose. Once you compassionately let the camel stick his nose inside the tent, how can you refuse him further entrance?

The problem of the deceased wife's sister is no longer with us, but the roadblock created by the tale of the Camel's Nose still is. The punch line goes by many names, among which are the Thin Edge of the Wedge and the Slippery Slope. The idea is always the same: we cannot budge a millimeter from our present position without sliding all the way to Hell. As with the Delaney Amendment, the fear of the Nose/Wedge/Slope is rooted in thinking that is wholly literate and adamantly antinumerate.

In the present century opposition to contraception has played much the same role that opposition to liberalization of the marriage laws did in England in the last century. The rhetoric has been very similar. Norman St. John-Stevas, a Catholic Member of Parliament, defended the Vatican's position of prohibiting all forms of contraception by asserting that "if any variation is admitted in *coitus* of a fundamental nature, unlimited variations must be allowed." In other words, The whole Camel: all the way to Hell.

In the matter of marriage to deceased wife's sister, Parliament took 125 years to decree the obvious. How long will the Pope take?

Like a skeleton key, the Slippery Slope argument fits a multitude of cases. We note with amusement a letter of Richard Nixon's published in *The New York Times* on 29 October 1965. At this time he held no public office. His letter warned that victory for the Viet Cong (the Communist-supported faction in Vietnam)

> would mean ultimately *the destruction of freedom of speech for all men for all time* not only in Asia but the United States as well. . . . We must never forget that if the war in Vietnam is lost . . . the right of free speech will be extinguished throughout the world.

I trust I may be forgiven for italicizing the most significant words in this passage. Mr. Nixon's subsequent good fortune after we lost the war in Vietnam sufficiently refuted his thesis. A decade after the debacle in Southeast Asia, a decade after the humiliation of the President in the White House, Mr. Nixon is still speaking freely and making a mint of money doing so. If there was a Slippery Slope, for Mr. Nixon it sloped another way.

The bioethicist Joseph Fletcher has astutely uncovered the implications of the Nose/Wedge/Slope argument:

> The wedge objection *cuts both ways:* If we refuse to do a thing which bears the possibility of abuse for fear we will find it easier and easier to tolerate the evil, then we will by the same token find it easier and easier to tolerate [having] less of the good. The wedge argument has no place in standard works on philosophical ethics or moral theology. The classical moralists' reply to it has been *abusus non tollit usum,* the abuse of a thing does not bar its use.

Those who take the wedge argument with the utmost seriousness act as though they think human beings are completely devoid of practical judgment. Countless examples from everyday life show the pessimists are wrong. Take the matter of automobile speed. Every now and then the newspapers inform us that a toddler has been killed by the family automobile, which, with the brake poorly set, rolled down the driveway and ran over the child. What was the speed of the car? Perhaps two miles an hour; perhaps less. Obviously even a very low speed may kill now and then. If we took the wedge argument utterly seriously, we would pass a law forbidding all vehicles to travel at any speed greater than zero. That would be an easy way out of the moral problem. But we pass no such law.

Instead we pass a law that sets an arbitrary line considerably above zero—55 miles per hour, 65, 75, or whatever. At the arbitrary line, no matter where it is, there will be some cheating, but a little cheating is better than no decision. By our actions we indicate that we don't buy the Nose/Wedge/Slope argument. We are willing to take risks, we are willing to draw arbitrary lines, and we are willing to be tough enough to enforce arbitrary laws. *Abusus non tollit usum.*

The logic that applies to admittedly ethical dilemmas applies also to environmental ones. In seeking a consensus, our first step should be to grant a modicum of justification for the position of society's nay-sayers. For the sake of argument we may grant that:

- Forbidding the use of boys to clean chimneys may somewhat weaken the spirit of commercial enterprise.
- Fitting soldiers with differentiated left and right boots may remove the character-building influence of pain.
- Painting Plimsoll lines on ships creates a bureaucracy that may, in time, become overbearing.
- Letting a man marry his deceased wife's sister may encourage murder.
- Permitting contraception may encourage sexual promiscuity.
- Allowing farmers to use herbicides and pesticides may result

in the deaths of birds and the development of chemical-resistant strains of pests.
- Putting chemical preservatives into food may increase the number of cancer deaths in later years.
- Allowing any air pollution greater than zero may have adverse effects on human health.

The radical environmentalist's "Everything is connected to everything else" is blood-brother to the Nose/Wedge/Slope argument: both are ecolate in a literate fashion only, not numerately so. Excessive ecolacy can lead to conservatism of the most stultifying sort. For prudence's sake, ecolacy must be combined with numeracy. Any action that we take—and inaction is a form of action—leads to some unwanted consequences. Prudence dictates that we compare the advantages and disadvantages of all proposed courses of action, choosing the one that, on balance, is quantitatively best.

Ecolacy vs. the Idea of Progress

The idea of Progress has ancient roots, but it came to full flower only after the Marquis de Condorcet courageously bore witness to it just before he was killed in the French Revolution. His book, published posthumously, bears the English title *An Historical Picture of the Progress of the Human Mind.* The author admirably summarized his conclusion in a few words: "Nature has assigned no limit to the perfecting of the human faculties . . . the perfectibility of man is truly indefinite."

True? There's no telling until we come up against some limit that can be proven to be a final limit. This may or may not come to pass. In the meantime, true or not, Condorcet's thesis is immensely inspiring. It promotes action. It breeds optimism.

If any change in intellectual climate deserves the accolade of "revolution," the worldwide diffusion of the idea of Progress does. It takes considerable historical research, and even more imagination, to appreciate what a different world men and women lived in before the idea of Progress shaped their unconscious expectations.

In achieving its dominion every revolutionary idea changes and becomes corrupted to a greater or less extent. For the idea of Progress, the corruption consisted in replacing Condorcet's broad vision of the progressive growth of the faculties of the human mind with a focus on the artefacts produced by that mind—the machines, the constructions, the physical inventions. By the twentieth century it is fair to say that Progress, for the vast majority of people in the Western world, had come to be identified with two items of faith:

Anything we can dream of, we can invent.
Anything we can invent, we are required to use.

He who clings to any faith will be disturbed by a resolute questioning of the foundations of his faith. Attempts to establish facilities for the testing of drugs and consumer products, as well as attempts to carry out "technology assessment" before a new technology is vested with power, have been continuously harried by the "true believers" of Progress as well as by the guerrilla forces of grasping commercial enterprisers.

There is a kernel of justification for the opposition to assays-before-action. Any government agency for testing or assaying, or for publicizing the results of tests or assays, must be staffed by custodians who, ideally, should not be answerable to commercial interests. But individuals may be moved by self-interest of many sorts; the profit-seeking interest of enterprisers is matched by the empire-building interest of bureaucrats. We must never forget the millennia-old question *"Quis custodiet ipsos custodes?"*—"Who shall watch the watchers themselves?" Custodians need watching, too.

Progress is a god to some people; so also is "the Environment" to others. When worshippers of the opposing gods

Environment and Progress meet, sparks fly. But not much light results from the encounter.

Environment and Progress can be reconciled, but only if enough passion is removed from each extreme to justify changing the initial letters to lower case. We can do very well without more capital letters or new religions. We must return to Condorcet's vision of progress as a process that takes place primarily in the mental realm. Technology is at best a reflection of the human mind. Ecology and the environment are mental constructs. The ecolate filter is a mental filter. We can be ruined if ecology is turned into a religion.

Semantic Aspects of Ecology

An important part of the ecolate filtering apparatus is semantic. Many conflicts over environmental matters can be cleared up quickly once the semantics are straightened out. Traditional language often throws a smokescreen over problems. A few instances:

- "We can't afford to spend much time worrying about waste." But *waste is a mental construct, not an objective reality.* The products of an industrial process are just products, period. Some of the products have a ready market. The manufacturer is well advised to seek out markets for the "by-products," or— that failing— convert them to other products that he can sell. Calling a product "waste" inhibits creative thought.
- "It's just a side effect." This is not objective description: it is interpretation biased by desire—the desire that some of the consequences would just go away and leave us alone. Objectively speaking, *there are no side effects.*
- "We might as well throw it away." But *there is no away to throw to.* Every single thing, including all the things we call "waste," is necessarily in our world. Sooner or later we will encounter our wastes again.
- "We need pesticides to get rid of pests." Quite so: but again we are using language in an attempt to control reality. *All so-*

called "pesticides" are, in reality, biocides, substances that kill life. We must ask, What kinds of life are killed, and at what cost? The conventional word acts as a blinder to the mind's eye. The so-called side effects of pesticides came as unpleasant surprises because the word "pesticide" was not challenged until Rachel Carson came along.

Whatever plan of action we adopt in our attempt to remake the world, our usual first step is to pin a laudatory label on what we are doing. We may call it development, cure, correction, improvement, help, or progress. We load untested conclusions onto ill-stated premises. But *every intervention in an existing system is, for certain, only an intervention.* We will make progress faster if we honestly call the changes "interventions" only, until an audit shows what we have actually done. Needless to say, such honesty will be resisted by most promoters of change.

Significant interventions in the world are always interventions in systems,of which three kinds can be usefully distinguished:

Organisms. An organism, be it bug or man, is a fantastically complicated system of cells, tissues, organs, and organ systems. Three centuries of medical progress have taught us much about these systems, but much more remains to be learned. In the meantime our distress at witnessing the malfunctioning of human organisms makes us willing to risk intervening medically with chemicals, both natural and manmade. Every intervention carries a risk, but by honestly auditing the outcome of our interventions, we make progress in medicine.

Systems of organisms. However much human beings may seek to make themselves independent of other organisms, our well-being depends on the survival in good health of many other kinds of plants, animals, and microorganisms. Except for those who are far gone in the worship of technology, most people want to live in a world richly populated by other species—birds, flowers, mammals, reptiles, ferns, mosses, predators, and prey. Massive technological interventions in the environment can bring about the massive extinction of species. Preventing this will not be easy. The attempt strains our knowledge and our political abilities to the limit. But we are making considerable progress in understanding—and a little bit of progress in applying our understanding.

Political systems. These systems are found only in the species *Homo sapiens*—"the political animal," as Aristotle called our species. Political systems are the least understood of all, despite the fact that great minds have puzzled about them for more than two thousand years. Why does an apparently well-functioning nation eventually become extinct? Is every system of political governance inherently unstable? Will self-interest suffice as the single motive power of a working system? If not, what else is required? Is religion (or something like it) required? Can a stable system be made up of rational elements only? Can we survive in the face of our worst impulses? Can we survive the sometimes disconcerting consequences of indulging our best impulses?

Our knowledge of the statics and dynamics of political systems is surely no more advanced at this time than was the knowledge of the human body before William Harvey. For humanity this is a pity—but what a magnificent opportunity this primitive state of political knowledge offers to the yet unidentified Harveys of political science!

There are still worlds to be discovered.

CHAPTER 8

The Binding of Time

If etymology were all that mattered, economics and ecology should be as cooperative as brothers, for their names are both derived from the Greek root *oikos,* meaning home or household. The first discipline deals with the accounts of the household of human beings, while the second ranges over all living things. Even when the ecological focus is on man (as it is in "human ecology"), the bearing of other living things on the welfare of human beings is not neglected (as it all too often is in economics).

As an academic subject, economics (first called "political economy") dates from the first half of the nineteenth century, ecology from the last half. Growing up independently the two subjects developed sharp differences. (Isolation during biological evolution also causes animal populations to diverge in character.) By the time ecology came to the notice of the general public in the 1960s, the language and assumptions of the two disciplines were strikingly different. Antagonism characterized many of the early encounters of economists and ecologists in the public forum. Nowhere was this more noticeable than in the way the two kinds of specialists look at the future.

Discounting the Future

In the early 1970s the governors of the Rockefeller Foundation and the Conservation Foundation had a try at closing the gap between economists and ecologists. A conference was convened in Washington, D.C., with about ten members from each camp, together with an equal number of foundation personnel. I was one of the ecologists.

The first day seemed a waste of time: par for the course. The behavior of academics on such occasions is more animal than intellectual. We were like a bunch of strange dogs, circling stiff-legged around each other, sizing each other up. Few issues were joined that day: it was a time for parry and thrust.

It was well into the second day before we made any real progress. We found that one element more than all others divided us: our conception of time as an element binding human decisions. What is the dividing line between "near term" and "far term"? For the economists it was almost always five years; for the ecologists (biologists all) it was more variable, but it was likely to be a century or more. ("The present" in the sister science of geology includes anything that has happened in the last million years.) It was only after we brought this difference out into the open that the discussion made progress. Significant differences still existed, but at least we now understood where the differences lay.

To say that economics is near-sighted is not necessarily to condemn it. The near-sightedness is understandable. Most of the time economics is a handmaiden of business, and experience has shown that it is very difficult to predict what business will be like more than five years in the future. Within this limited time frame, however, economics has developed a persuasive way of dealing with time on what appears to be a wholly rational basis. This way is called "discounting the future." It is a way of binding time to value.

Suppose I am offered a choice of two goods, $100 today, or $100 a year from now: which should I prefer? Under most circumstances the rational thing to do is to choose $100 now.

What about $100 now versus $1,000,000 a year from now? Unless one suspects some very peculiar hidden conditions, the rational action is to choose the distant good in this case. (In developing the economic concept of rationality we first ignore such complications as the reliability of promises.)

Rationality tells us there must be a dividing line between

the present-good preference and the distant-good preference, some defensible justification of our method of choosing from among time-bound choices. The economist's explanation runs as follows.

The key element is the rate at which interest is earned on invested money. Suppose the present rate is 10 percent per year. In the first situation described, I can elect to take $100-now, and then put the money out at interest, which will yield me $110-a-year-from-now, which is certainly preferable to a delayed payment of only $100. We call this preference rational.

The rational dividing line between choices that are time-bound and purely economic is, then, set by the rate of interest. We say that $110-a-year-from-now must be discounted to $100-now. Economists also call $100-now "the present value" of $110-a-year-from-now (given an interest rate of 10 percent).

It is normal business practice to compound interest at regular intervals (or "instantaneously," as can now be easily done using the exponential functions wired into inexpensive hand calculators). The potential exponential growth of money has the effect of discounting the distant future severely. For instance, since $100 at 10 percent interest would grow to $1,083.47 in 25 years, we would be monumentally foolish to choose $100-twenty-five-years-from-now to $100-now.

To put the matter another way, assuming a trustworthy interest rate of 10 percent per annum, the present value of $100-twenty-five-years-from-now is only $9.23, for the sum of $9.23 put out at 10 percent interest will grow to $100 in 25 years. The rational man should not pay any sum greater than $9.23 for the right to receive $100 a quarter of a century from now.

Who Can Afford a Forest?

Table 8-1 gives the present value, in dollars, of one million dollars at some later time, assuming various interest rates and various periods of time. The table can be used to determine

TABLE 8-1
Table of Present Values

The "present value" of $1,000,000 at a future time
(given various interest rates, for varying periods of time)

Interest rate per year	**Time in years**					
	5	10	20	50	100	500
3%	$862,609	$744,094	$553,676	$228,107	$52,033	38 cents
6%	747,258	558,395	311,805	54,289	2,947	less than a penny
10%	620,921	385,543	148,644	8,519	73	less than a penny
20%	401,878	161,506	26,084	110	one penny	less than a penny
50%	131,687	17,342	301	less than a penny	less than a penny	less than a penny

the advisability of possible investments. Suppose, for example, we are considering planting a forest that takes 20 years to mature and yield an income. Expenses (taxes, fire protection, et cetera) are incurred during all this period, but for simplicity, let us assume that all the expenses occur in Year 1, and all the income is gained in Year 20. How much can we afford to spend to gain a gross income of $1,000,000 twenty years from now?

In the real world, of course, our calculation must take account of additional complications. Will there be inflation, i.e., how much will $1,000,000 be worth 20 years from now? Will the interest rate go up or down, and by how much? Will the tax laws be amended in ways that are favorable or unfavorable to this enterprise? For the present we will ignore all such questions to get at the essence of time-bound decisions.

From the table we see that we can afford to pay no more than $148,644 now to gain a million dollars 20 years from now, if 10 percent is the going rate of interest. If such an expenditure promises to yield less than a million dollars later, it is not rational to undertake the enterprise. (If the prevailing rate of interest were lower it would be economically rational to spend more, of course. At 6 percent, an expenditure of $311,805 would be rational.)

What kind of a forest could we get for our money? Probably one of "pulpwoods," that is, trees that grow rapidly and can be converted into paper. We could also grow for "biomass," that is, for wood that can be burned for heat or converted into a variety of chemical products. But we probably will avoid growing hardwood trees because these take a long time to mature, which means that we have to look at the later columns of the table. Note that at a 20 percent interest rate (which prevailed for a while for commercial loans in the early 1980s), if the forest took a full century to mature, the present value of a millon-dollar harvest is almost exactly one cent. *At high rates of interest the present value of the distant future effectively vanishes.*

Such analysis leads biologists to wonder if "the rational man" can ever afford to plant hardwood forests? Economists, with more faith in technology, suggest that scientific advances in silviculture will find ways of making hardwoods grow faster. No doubt: but faster growth means wider growth rings, which is just what we don't want in a quality hardwood that is to be used for fine furniture or good hardwood floors. The best hardwood comes from trees that have grown slowly under difficult conditions. Fast growth and high quality are incompatible. Stalemate.

Not so, says the technological optimist. If all we want is something hard, we can grind up various trees to make a hard "chipboard." Quite so: but suppose we want a beautiful grain, too? Simple, is the reply: take a few choice trees and (using a lathe) shave off a microscopically thin layer and use this as a veneer on chipboard. Yes: but suppose your fingernail scratches the veneer, then what have you got? . . . And so the argument goes on between those who support the purest economic rationalism and those who wonder if a world that is so simply rational is the kind of world we want to live in.

The hardwood forest is, of course, only one of many examples of natural goods that take a long time to produce. We could, if we were foolish enough, cut down all our great redwood forests, and then plant them again. Redwoods grow very fast, but the esthetic entity we call a redwood forest is a complex mixture of many species of plants of different ages dominated by gargantuan trees whose origins stretch back to before the birth of Christ. He who finds ecstasy in the wonder of today's mature redwood forest benefits from a preservation a pre-Christian economist could not have justified. No matter what ultimate value we assign to the mature forest, at any believable rate of interest the economist's "present value" of a good that lies 2,000 years in the future is practically indistinguishable from zero. If we are content to act as the pure "economic man," we may conclude that we can't afford such a good. But does it not seem odd that we, the inhabitants of one of the richest countries in the world, cannot justify the best? Pure economic rationality does not yield an acceptable answer; we need to find a higher form of rationality.

Forests that are difficult to justify in a private-enterprise setting are even harder to justify when the gains from forest growth are diffused over a large population. This is true despite the fact that many people (though perhaps much less than a majority) understand the importance of forests in controlling floods in a river drainage system. Forested areas act like sponges, soaking up the too abundant water of a rainstorm, releasing it later over an extended period of time. The water oozing out of the sponge of a forest is clean water; that which comes from a deforested area is murky, laden with eroded soil.

Under conditions of high population, pressures toward deforestation are likely to be irresistible. The short-term gain from cutting down the trees overwhelms the perceived long-term gain from leaving them in place to protect the soil and prevent floods. The long-term gain accrues to people distant both in time and space from those who stand to gain by cutting down trees here and now. The eventual consequences of short-term thinking can be devastating, as they were in China in 1931. Because of extensive deforestation in the headwaters of the Yangtze River in earlier centuries, heavy rains in 1931 produced floods in the river basin that killed 3 million people and dispossessed more than 40 million.

In response to this disaster, postrevolutionary China has reforested nearly 200 million acres, and the cure has worked. Just fifty years after the disastrous flood of 1931, an even greater rainfall was experienced, but the losses in life and property were less than one-thousandth what they were earlier.

China has learned her lesson, but have we learned ours? Under the direction of the Army Corps of Engineers, we continue to try to control floods by the brute force of levees and dams in the lowlands rather than by reforesting the distant highlands where floods are generated. Levees, like cocaine, are addictive: building them increases the craving for more. Raise the levees and you raise the bed of the river as more silt is deposited. This in turn requires the levees to be raised still further, thus increasing the damage done whenever, after many such cycles, the river finally breaks free of man-made controls. St. Louis and New Orleans are living on borrowed time, and all because the simple-minded policy of controlling floods by building levees and dams has not been challenged with the ecolate question, "And then what?"

The Rationale of Poverty

Numeracy is at the heart of economic rationality, but in arriving at rational decisions, exact accounting is often less important than being guided by the *spirit* of numeracy. Long before either literacy or numeracy were well developed, human beings must have been making decisions of the sort that would be dictated by a table like the one presented here. Rational but unschooled human beings all over the world act this way. The human brain is like a hidden computer that, unbeknownst to us, grinds out silent numbers on which our decisions are based. By its nature this hypothesis cannot be proved, but many phenomena are consistent with it.

A billion or more people scattered throughout the world are now living in what the social anthropologist Oscar Lewis called "the culture of poverty." Most of these people are illiterate; on the surface, they may appear to be equally innumerate. A striking characteristic of the culture of poverty is its invariable "present-orientation." Effective concern for the future is almost entirely missing among the desperately poor.

Such present-orientation is easy to understand. It is as though poor peasants were making decisions using a hidden Table of Present Values, in which the possible interest rates are all very high. This is more than a gratuitous assumption. It is characteristic of poor countries that the interest charged by money lenders is shockingly high by our standards. Interest rates in excess of 50 percent per annum are not uncommon.

High rates of interest were also common in the Western world when it was no richer than what we now call the Third World. In the fourteenth century Frankfurt put out the welcome mat for Jewish bankers who would promise to charge no more than 32.5 percent interest. (Christian moneylenders were then charging up to 266 percent.) The future-oriented planning so characteristic of capitalism could not get started in the West until interest rates came down drastically.

Looking at a culture of poverty from our favored position in a prosperous society, we are tempted to express moral indignation at moneylenders' treatment of the poor. Yet within the framework of poverty their behavior is completely rational. Those who want to borrow are, by legitimate banking standards, poor risks. A lender must charge high interest to borrowers so that the few who repay can keep him from being ruined by the many who default. Looking at the transaction from the other side, the present need of the borrower is so great that he is willing to forego a great deal of an uncertain future for a certain present gain. Moreover, every man knows that it is always possible that he may die before tomorrow arrives. This knowledge is particularly poignant in poor societies.

A culture of poverty is one in which the future is discounted—both implicitly and explicitly—at a very high rate.

This way of looking at societies also explains why it is that relative differences in wealth are greatest in poor societies. (It is one of the odd characteristics of political radicals in our society that they think that relative differences in wealth are greatest in the wealthiest societies. This is not so. Why do radicals cling to such an erroneous belief? I submit the following as an explanation. This belief increases the ability of radicals to evoke guilt feelings among their prosperous fellow citizens, thus preparing the ground for their radical proposals to redistribute the wealth.)

It is necessarily true that a wretchedly poor society generates the greatest relative differences in the distribution of wealth. This can be shown by a simple thought-experiment. Imagine an initial situation in which there were absolutely no differences in wealth. If by chance even the slightest difference appeared, this difference would escalate, for the following reasons.

We can think of all individuals in a given society as unconsciously consulting the same hidden Table of Present Values (like our Table 8-1); however, each individual can "see" only one line, corresponding to his personal rate of discount. A rich man may use the 3 percent line, a poor man the 20 percent. Offered the same opportunity the two quite properly reach different decisions. Both are economic rationalists, but—like the two bishops on the same side in a chess game—their minds follow different paths. The same economic calculus operating against different backgrounds of wealth responds differently to the "same" business opportunity.

As a result of their different responses (both rational), when the future arrives the rich man has climbed one additional step up the ladder of wealth. The relative difference between the rich man's and the poor man's wealth has increased. Now that he is wealthier, the rich man can afford to discount the future at a still lower rate, thus further increasing his advantage in bargaining with his poor compatriots. Thus does rational behavior sustain and augment differences in wealth.

The runaway process that accounts for the unequal distribution of wealth in society is called "positive feedback" by engineers. The higher the rate of interest at which the poor discount the future (relative to the rate assumed by the rich), the surer it is that greater differences in the distribution of wealth will develop. Of course, other processes are also brought into play in poor societies. Most noteworthy is the corruption of the government by the wealthy class. Such additional factors augment the effect of the primary one.

Some people are affronted by this way of looking at poverty, regarding it as a rationalization by selfish people to justify hanging on to their wealth. Well-meaning reformers are not inclined to quote *Matthew* 26:11, wherein Jesus said, "For ye have the poor always with you." Nor do they expatiate on the preceding chapter of the same book, wherein there is a most remarkable eulogy of capitalism. In our day there are a thousand who profess reverence for the Bible for each one who reads it thoughtfully.

A successful policy for the mitigation of poverty must take account of the mechanisms that create income differences. Traditional economic systems, left to themselves, spontaneously create and sustain differences. Inequality of wealth is, as it were, the entropic state; to create (and sustain) any other distribution requires "energy" of some sort. At the very least an acceptable distribution requires continual vigilance. Equidistribution is inherently an unstable equilibrium.

Though we generally seek *stable* equilibria, there are situations in which only unstable equilibria are possible. The flight of a hummingbird, unlike that of a glider, is perpetually in unstable equilibrium; yet the hummingbird manages to fly very well. But whereas a man in a glider could afford to let his attention wander as he admires the scenery, the sensory reactive system of the bird dares not relax its vigilance even for a moment. So it would be with any society that made equidistribution of wealth a primary goal: only untiring vigilance can maintain such an improbable state—and the "cost," in some sense, would be high.

Because the spontaneous pressure toward inequality is greatest in societies in which the average income is lowest, those who have the strongest desire to create a society of economic equals should recognize that their first goal should be to keep the population at a low enough level to insure a high average income. This will make it easier to institute other measures. Few enthusiasts for "economic justice" realize that overpopulation favors economic inequality. The redistribution of wealth sought by such reformers would increase population and thus secondarily increase inequalities of wealth. Like the generality of reformers, they fail to ask the ecolate question "And then what?" before proposing their reforms. No amount of good intentions can excuse a counterproductive measure.

Inflation

The ever-present danger of inflation would be much less if every generation could more easily assimilate the feelings of those who have in the past gone through the harrowing experience of runaway inflation. Perhaps a vignette from Germany in 1923 can help activate the imagination.

> Two women on their way to the bank with a laundry basket full of paper money pass a shop window around which many people are gathered. Wanting to know what the attraction is, they set their basket down and push through the crowd. When they have satisfied themselves that the attraction is of no interest to them they turn around to pick up their load. To their chagrin they find that someone has emptied the money out on the sidewalk and fled with the basket.

This account neatly epitomizes the evil of inflation at its most extreme—"hyperinflation," as it is called. Money is not a good in itself; rather it is a somewhat generalized pledge against the future. Inflation discounts the future, forcing the rational person to choose present real goods (a basket) over the promise of goods in the future (paper money). With hyperinflation the "present value" of the future soon sinks to zero. Before the 1923 runaway process was stopped, prices in Germany had risen 1,300,000 times. Hyperinflation in Hungary in 1945–46 was yet a million times worse. Under such conditions the normal rhythms of life are grossly disturbed, as is illustrated by another account from Germany.

> At about 11 o'clock the siren sounded, and from all over the factory thousands of people streamed into the forecourt where a five-ton truck was standing, brimful with paper money. Onto it climbed the chief cashier and his assistants. They read out names and just threw out great bundles of notes. As soon as you had caught one, you made a dash for the nearest shop and bought anything you could lay your hands on. This was essential, because at noon every day the new exchange rate came out, and if, by then, you had not converted your paper money into goods, you stood to lose a third or more of your salary.

The consequences of hyperinflation beautifully illustrate the meaning of the First Law of Ecology. A government that is unwilling or unable to stop the escalation of inflation does more than merely change the price of things; it turns loose a cascade of consequences the effects of which reach far into the future. Prudent citizens who have saved their money in bank accounts and government bonds are ruined. In times of inflation people spend wildly with little care for value, because the choice and price of an object are less important than that one put his money into material things. Fatalism takes over as society sinks down into a culture of poverty.

Wage conflicts become ever more bitter. Farmers are reluctant to sell their grain; city dwellers, in desperation, pillage the farms. Sharp dealing of all sorts becomes the rule. Retired couples on fixed incomes are impoverished; many commit suicide. People in occupations that have little bargaining power, e.g., teachers and writers, become disaffected. As an understandable result, literature is enlisted in the service of chaos: intellectuals preach disaffection, distrust of government, cynicism, fatalism, and despair of the future. Wordsmiths become the willing servants of revolution.

As the greatest evil of all, negative attitudes become the norm among the young, thus setting the stage for unpredictable disasters in the future. The principal unexpected development in Germany following the 1923 inflation was the rise of the Nazi party to power ten years later—with further grave consequences that cannot even yet be said to have worked themselves out. *We can never do merely one thing. . . .*

Politicians and the Future

The television commentator Bill Moyers, who was a member of the inner group of advisers to President Johnson, has commented tellingly on the outlook in the White House during the military buildup in Vietnam. The administration was immersed in a climate of opinion that led it to step deeper and deeper into the quagmire of Southeast Asia at a time when any minimally objective observer could confidently predict disaster. "With but rare exceptions," wrote Moyers, "we always seemed to be calculating the short-term consequences of each alternative at every step of the [policy-making] process, but not the long-term consequences. And with each succeeding short-range consequence we became more deeply a prisoner of the process."

The same process leads to escalating inflation, for the immediate consequences are often pleasant to many of those in power. Businessmen can make a profit on their inventories, union leaders can negotiate raises for their members, and the government can collect higher taxes and erase part of the national debt incurred by unwise decisions made earlier. Inflation is a marvelously covert form of taxation.

Why is it so difficult for politicians to take the long view? It is often said that the trouble is rooted in democracy, the argument running as follows. Conscientious politicians want to do a good job legislating and administrating; but all politicians, conscientious or not, have to be concerned with being reelected. Many political decisions that would make matters better in the distant future incur "unacceptable" costs in the present. Voters, most of whom have only a poor grasp of future necessities, will throw out an elected representative if he votes to increase taxes now. Noting that the term of representatives is only two years, of senators, six, and of the President, four, we may take as a rule of thumb that the horizon of most politicians is no more than five years away. Proximate goals drive out distant goals.

Blaming democracy for short-sightness is tempting, but the truth surely is not so simple. American businesses, which certainly are not run democratically, often suffer equally from short-sightedness. Back in the 1920s Henry Ford, a very shrewd businessman most of the time, lost millions of dollars and his position as the number-one car-maker when he resolutely closed his eyes to the commercial necessity of designing a new car to replace the Model T. In the 1960s the entire American automobile industry, given ample warning that consumer taste was moving toward small cars, buried its corporate head in the sand until forced to take this inevitable step at far greater expense than would have been required had it acted earlier.

And up until the very day that the oil-producing countries of the Near East tripled the price of oil, most of the American public acted on the assumption that oil would be cheap forever. Why such folly?

The answer is to be found not in broad abstractions like "democracy" or "stupidity" but in the particularities of everyday operations. The top executives of most businesses are strongly motivated by the award of year-end bonuses tied to profits made during the past twelve months. Decisions that promise greater gains several years from the present, provided heavy expenditures are made now, do not appeal to managers for two reasons. First, of course, is the usual and rational practice of discounting the future. Equally important is a second factor that is tied to the changing stages of the human life cycle.

By the time a person has been promoted to the level at which he is eligible to receive bonuses based on performance, he is also near the age of retirement. Typically, he is about five years away from the time when he will receive no more yearly bonuses. Given the choice of a decision that will probably increase this year's "bottom line" and one that *may* increase the bottom line during his successor's reign, it is an unusual executive who will prefer a possible distant good to a more probable present good. As a result the average businessman may have as restricted a time horizon as the average politician. In both instances the proximate goal tyrannizes over the distant.

At this point someone might suggest that perhaps we are dealing with an imperfection of America, or of the Western world generally? The rise to popularity of anthropology in the twentieth century has made people keenly aware of the dangers of ethnocentrism, of looking with scorn on all societies but one's own. We need to become equally aware of the shortcomings of the opposite attitude, which we might call "ethnofugalism." It is fashionable in some intellectual circles to assume gratuitously that other nations, other cultures, are superior to ours. Unprejudiced observation, however, shows that Americans are not alone in being imperfect.

The mobilization of community efforts in Russia, China, and India has standardized around the Five-Year Plan. Apparently it is not possible to persuade citizens to sacrifice much for more distant goals. It looks as though the five-year horizon is tied to properties of human nature that are more fundamental than the particular form of political or industrial organization. We need to know more about human nature. If the progress of scientific knowledge enables us to make reliable predictions of preventable catastrophes more than five years in the future, it is essential that we learn how to increase humanity's future orientation beyond the usual five-year horizon.

The Limits of Rationality

Rationalism is an ideal and a program. Programmatically, the hope is to put all human behavior on a rational basis, leaving no room for loose ends, mysticism, or the "No Entry" signs set out by theologians. At no time is the ideal fully achieved. In justifying his actions the rationalist frequently has to admit that he cannot do better than quote that tortured mystic Blaise Pascal: "The Heart has its reasons which Reason knows not of." The rationalist hopes that any surrender of the ideal of total rationalism is only temporary.

As concerns the economic binding of time to decisions, some rationalists are not yet entirely satisfied with the results. Consider the implications of the survival of two species of trees in China, the ginkgo and the dawn redwood. Both kinds were found (by explorers from the West) in temple courtyards, being nearly extinct elsewhere in China. This simple fact of distribution is highly significant.

The geographical areas in which the trees were found were severely deforested centuries, or even millennia, ago. The demands of population for building materials and fuel for the cooking of rice so diminished the "present value" of forests-yet-to-be that regrowth was simply not possible. Yet isolated trees were allowed to flourish unharmed in temple courtyards. For what reason?

In places where supplies of fuel were very short, one can imagine peasants coming into a temple courtyard in the hope of snipping off branches to use as fuel for cooking their rice. (We see such a process taking place on public roads in India today.) Peasants can soon destroy a tree in this way. The rational calculus of each peasant tells him that no economic "present value" of tomorrow's tree can equal the simple biological present value of his life. But, since the ginkgo and the dawn redwood did survive in temple courtyards during times of desperate poverty, something other than a rational economic calculus must have protected them. The caretaker-priest must have given a different answer, an answer couched in terms of the word "sacred."

Operationally, the word "sacred" falls into the same category as the word "infinite," which was discussed in Chapter 4: it is a signal that discussion is being peremptorily closed. To accept the authority of the sacred is to step outside the bounds of rationality. This is one way to escape the limitations imposed by the economic theory of discounting.

The temple trees were sacred, untouchable. It is common in all societies to grant a measure of sacredness to the artifacts of sacred places. If this explanation of the survival of rare trees in deforested areas is true, a remarkable weight must have been given to the concept of the sacred in old China. In the present heyday of Western individualism this may seem too much to postulate, but we should not overlook how often the king's game was protected against starving peasants in medieval Europe. To be the king's own was also to be sacred, and hence beyond the reach of ordinary economic reasoning as we conceive it in academic quarters today.

In a world ruled by the unconscious assumptions of extreme individualism, it is difficult to give a solely rational justification for saving things for another generation. It is not easy to answer this double question: "Why should I do anything for posterity? What has posterity done for me?" The generation that makes the sacrifice is not the generation that reaps the harvest.

The great eighteenth-century conservative Edmund Burke sidestepped the question with this observation: "People will not look forward to posterity who never look backward to their ancestors."

The hypothesis implied by Burke is a psychological one. It may well be true. If so, it suggests another escape from the unwanted consequences of the rational treatment of the future: tradition. An individual who *identifies* with his ancestors steps outside the bounds of pure individualism. Once this step is taken, identification with posterity and its interests is not so difficult. In fact, what we perceive as the limitations of the economic theory of "discounting the future" are perhaps merely the limitations of the concept of individualism, an ideology that has dominated Western society for several centuries now.

Burke's remark is the sort that one would expect from a securely situated landowner who is keenly aware that his ancestors passed their estate on to him. Preservation of the estate intact in the face of temptations to make present gains at the expense of the future comes easier to those who are wealthy. This, of course, fits in with the economic theory of poverty, in which motivation is proportional to the psychological "weight" of a gain or a loss—"marginal utility," economists call it—rather than to the absolute value (dollars or pounds). It is quite rational for an individual to base his decisions not on a situation-independent measure of money, but on its very personal marginal utility *to him*.

Often a psychic gain is experienced by sacrificing a present good for the sake of the future, e.g., the praise of one's associates for having performed an "unselfish" act. Unlike the marginal utility of money and material property, this secondary gain does not diminish with increasing wealth, and so there comes a point at which it psychologically pays a rich man to make decisions that are "unselfish" in the sense that they are posterity-oriented. Wealth lessens the burden of pure rationality, which is, after all, designed to deal with perceived shortages. No perceived shortage, no problem.

Many of the forests of Europe owe their persistence in the face of the demands of poverty to the fact that they were owned for centuries by privileged people—first the nobility, and later the merely wealthy. This remark could be criticized as a defense of privilege, but it is not. Moreover, there is another way to achieve the same end: public ownership. When ownership is spread over a large population, the cost to the individual (taxes) of preserving a forest or park is very little.

It would seem, therefore, that turning valuable resources into common property would be a good way to protect them. So it sometimes is, but commonizing property can also be a sure way of ensuring its destruction. We need now to explore the nature of common property, so we can understand when commonizing property is a good thing, and when a bad.

CHAPTER 9

A Tragic Distribution System

Space! After laying out thirty billion dollars to get to the moon, we spent a few billions more on space explorations. What did we get for our money? It would be short-sighted to say "Not much," for (as with Columbus' daring adventure) the unsought benefits may ultimately prove greater than the sought. Certain pictures taken of our globe from the vantage point of space have deepened our understanding of what is happening here on earth (though this is hardly what Congress had in mind when it voted funds to the space agency). One of the more significant of these space photographs was published and commented upon in the pages of the weekly journal *Science*, though it was little noticed elsewhere. It threw new light on a disaster that had upset many people.

In the early 1970s there was a prolonged drought in the Sahel, a broad band of semidesert extending across Central Africa and encroaching on several nations. At the height of the drought satellite pictures of Africa showed a puzzling green pentagon in the middle of the devastated region. When an agronomist at American University, Norman MacLeod, visited the site, he found a simple explanation: the area was surrounded by a barbed-wire fence. A 250,000 acre ranch, run as a private enterprise, was protected from the hordes of animals that were grazing the surrounding land to death—first the pasture's death, and then the animals'. The protected ranch was divided into five sections, only one section being grazed each year, with a rigid limitation on the size of the herd.

Space technology had given us a textbook illustration of a fundamental principle in political economy, for which I proposed the name "the tragedy of the commons" in 1968. The elements of this principle had been laid out in 1832 by William Forster Lloyd, a professor of political economy at Oxford. The memory of his contribution was almost entirely lost for more than a century.

A commons is a resource to which a population has free and unmanaged access; it contrasts with private property (accessible only to the owner) and with socialized property (access to which is controlled by managers appointed by some political unit). In the initial formulation the commons was a plot of land, but (as we shall find) it can be any resource subject to distribution among claimants. Since 1968 the theory of the commons has been fruitfully developed in economics, ecology, psychology, sociology, and many other fields.

Paid to Do Wrong

Contemporary Africa, rather than Lloyd's England, will be used to illustrate the properties of the commons. Outside the barbed-wire enclosure the land of the Sahel is regarded as common property by the nomads who live there. Each individual, each tribe, can graze as many animals on the commons as he (they) can lay hands on. Probably few of the nomads have heard of Karl Marx, but in exploiting the pasture they follow the principle Marx set forth in his "Critique of the Gotha Program" in 1875: *From each according to his ability, to each according to his needs!* The exclamation point is Marx's, and the importance he subscribed to this statement justifies our putting it in italics. Marx called this prescription his banner, and his followers took it seriously. So should we. Let us see what it means operationally, when applied to a situation like that of the Sahel.

In practice, "needs" are determined by each individual or tribe. Since herd animals are the basic form of local money, individuals and tribes tend to take a generous view of their own "needs."

Indefinite increase in the size of herds sooner or later produces a number of animals that is far beyond the biological "carrying capacity" of the pasture. Good grasses are selectively eaten, leaving less nourishing plants to take over. Ever increasing grazing pressure finally destroys even the less toothsome plants. Deprived of its normal plant cover, the soil is laid open to erosion by wind (and water, if there is any). Animals die by the tens of thousands, particularly in a drought year. Droughts are normal to this region: several occur every generation. To succeed, a policy for exploiting the resources of the Sahel must allow for recurrent and unpredictable droughts.

Human beings also die in vast numbers in a drought year, though the exact numbers in this uncensused land are subject to dispute. In recent years philanthropy-at-a-distance has averted some human deaths through food shipments to the suffering populations. Modern medicines have also saved lives. Temporary gains in food productivity have been achieved by substituting high-paying cash crops for subsistence agriculture, and by shortening traditional fallow periods on the grazing lands. Time has shown that these short-term gains were achieved at the expense of long-term losses.

The Sahelians saved by outside intervention in the 1970s continue to multiply both themselves and their herds, thus eroding the resource base of their life. The natives (through tradition) and foreigners (through philanthropy) have worked hand in hand to destroy the Sahelian environment. (But their intentions were good!)

Exact figures on earlier disasters are lacking because, until recent times, the world press did not pay much attention to what happened in the Sahel. Thousands of people could die there without their suffering making the front page of newspapers. Poverty is not a modern invention, but instantaneous worldwide publicity is.

The basic problem of the commons is (as we shall see)

psychological; the consequences are ethical, economic, and political. The theory is based on the concept of "the economic man," that is, a person whose major motive is self-interest. How should we expect such a person to behave when a resource is owned in common but the distribution of the benefits follows the Marxist rule, "to each according to his needs"?

Imagine yourself as a herdsman in the Sahel at a time when the total population of herd animals has just reached the carrying capacity of the land. Suppose you have a chance to acquire ten more animals. Suppose also that you are in complete possession of the facts—that you understand carrying capacity and the dangers of transgressing it. Should you, or should you not, add ten more animals to your herd?

Since the additional animals are (by hypothesis) ten more than the carrying capacity, all your animals will have a little less food per capita next year than this. So will everyone else's animals. Even so, you expect a net gain from the acquisition, for this reason: the gain is all yours, but the loss (from transgressing the carrying capacity) is shared among all the herdsmen. Your share of the loss is only a small fraction of the total. Balancing *your* gain against *your* loss you decide to take on ten more animals. In economics this is called a rational decision. To behave otherwise would be to behave irrationally—in the short run.

Every other herdsman in a commons must, if rational, reach the same decision—not only this year but in every succeeding year. In the long run this kind of behavior produces disaster for all, as overgrazing turns semidesert into desert. Even if you understand completely the disastrous consequences of living by the rules of the commons, you are unable to behave otherwise. The rules pay you to do the wrong thing.

As a good citizen you might refuse to add to your herd, but what makes you think every other herdsman would also be a good citizen? If even one participant in the commons should act in a "selfish" (read, "rational") way, your restraint would go for nought. As selfish and rational exploiters prosper at the expense of the public-spirited, envy will cause some of the latter to join the "rational" decision makers in their ruinous behavior. What might begin as the selfish rationalism of a few, ends in the corruption of the many.

Those who think a Marxist distribution system in a commons would work "if only people would act like good citizens" fail to realize that they are expecting the impossible, namely that every single person should be an angel. By contrast, those who hold out no hope for the Marxist system assume merely that every sizable population includes at least one person who is less than an angel. This is a modest assumption.

Additionally, the non-Marxist assumes that envy is part of the normal endowment of human beings. Profound religious thinkers have said as much for centuries; *invidia* is one of the Seven Deadly Sins. Nevertheless the theory of Marxism is built on the assumption that envy does not exist, though (paradoxically) the evangelism of Marxists is shot through with appeals to envy.

To repeat, the system of the commons ends in disaster even if every member understands the situation completely. This gives new meaning to the ancient idea of tragedy, which (to the Greeks) was a disaster that even foreknowledge could not prevent. The tragedy of the commons is a logical consequence of the rules of the game ("to each according to his needs") coupled with inescapable human nature (some people, at least, are both competitive and envious).

To escape the tragedy, the unmanaged commons must be abandoned. The community must reject the rule, "to each according to his needs." The Marxist assumption that "need creates right" must be flatly denied.

Responsibility: Sense and Nonsense

Any discussion of the proper relationship between the community and the individual sooner or later trips over the word "responsibility." The person using the word seldom defines it; rather, he or she employs seductive rhetoric to gain approval of actions contemplated or performed. Two examples from history and one from literature should make these points clear.

On the night of 30 June 1934, at the order of Der Fuehrer, Hitler's longtime supporter Ernst Roehm was killed, along with 100 to 1,000 of Roehm's gang. (Understandably, no one was standing around counting.) Two weeks later Adolf Hitler justified his action with these words:

> If anyone reproaches me and asks why I did not resort to the regular courts of justice, then all I can say is this: In this hour I was responsible for the fate of the German people, and thereby I became the supreme judge of the German people.

When Nicholas II, the last tsar of Russia, refused to put an end to the civil disabilities of the Jews, he justified his decision thus:

> Up to now my conscience has never led me astray, or been mistaken. On that account I am going, once more, to obey its dictates. I know that you, as well as myself, believe that the heart of the tsar is in the hands of God. Let it therefore remain so. I am bearing in the sight of the Almighty a terrible responsibility in regard to the power which I possess and wield, but I stand always ready to render Him an account.

In his novel *The Brothers Karamazov* Dostoevsky has Father Zossima say:

> You have only one means of salvation: take hold of yourself and make yourself responsible for all men's sins. My friend, believe me, that really is so, for the moment you make yourself responsible in all sincerity for everyone and everything, you will see at once that it really is so and that you are, in fact, responsible for everyone and everything.

What, in fact, have these three people said about responsibility? Essentially, nothing. They have not pledged themselves to correct the mistakes they make. To whom did they acknowledge responsibility? Hitler, to himself; the tsar, to God; and Father Zossima, to everyone (which is a sly way of saying, to no one). Like the proverbial diplomat, each man found exactly the right words with which to promise everything and guarantee nothing. In public matters the difference between promising and guaranteeing is the difference between saying with a great show of earnestness, "Cross my heart and hope to die," and swearing in court on pain of perjury.

The word "responsibility," if it is to be of any use in matters of public policy, must be employed only when it is clear who is answerable to whom. It must define a substantive relationship, not a mere feeling. I know of no finer operational definition of responsibility than the one given us by the philosopher Charles Frankel:

> *A decision is responsible when the man or group that makes it has to answer for it to those who are directly or indirectly affected by it.*

Those directly affected by Tsar Nicholas's decision were the Jews, with all Russia being indirectly affected; but Nicholas answered only to his God (not the Jews'). Those directly affected by Hitler's action numbered in the hundreds with all Germany (and ultimately, a substantial part of the world) being indirectly affected: but Hitler answered to no one. As for the words attributed to Father Zossima, are they more than mere eloquence?

Hortatory responsibility may play a useful role in personal relations; in public policy, only operational responsibility matters. The consequences of the various alternative systems of distribution must be sought in the kinds of operational responsibility they create.

Who Benefits? Who Pays?

Discussion of the distribution of goods and bads in society is too often unproductive because of excessive reliance on abstract nouns—democracy, communism, socialism, exploitation, rights, justice, and the like. These great abstractions have been used for so long by so many people with so many different axes to grind that their continued use leads to more *cant* than sense. (Pardon the use of this antique word, but it is one that should not be allowed to fade away: it is a modification of the word "chant," and aptly describes a discussion in which sacred words are chanted with little or no thought of their operational meaning. Most public discussions of "responsibility" are mere cant.)

To cut through the cant of "responsibility," we must ask the double question "Who benefits? Who pays?" This is the first question to ask when a politico-economic system of distribution is proposed. It focuses our attention on operations and their consequences rather than on words. The answer to this double question largely defines the properties of a system.

We take it as axiomatic that every social action entails both gain (profit) and cost (loss). We can indicate the way profit and loss are distributed by three alternative verbs: privatize, commonize and socialize. These verbs operationally define the three major distribution paths.

Privatize. Profit or loss is privatized when it accrues wholly to an individual (or, with less precision, to a collection of individuals called a legal person, e.g., a corporation).

Commonize. Gains or costs that are spread out indifferently over a whole population are said to be commonized. The "commons" of the Sahel, discussed above, is an "unmanaged commons," in contrast to the type to be mentioned next.

Socialize. When profits and losses are differentially distributed by managers (bureaucrats) among the group that owns the common property we say that the property is socialized. The system that does this may be called "socialism"; it is also called a "managed commons." A managed commons, though it may have other defects, is not automatically subject to the tragic fate of the unmanaged commons.

(In passing, note that the word "comm*u*nism" will not be used here because of its ambiguity. Sometimes it refers to socialism, sometimes to "commonism," a system in which losses are commonized, as defined above. In terms of their consequences socialism and commonism are radically different; we should not allow ourselves to be befuddled by the term "communism," hallowed though it be by usage.)

What we call private enterprise or "privatism" is a system which, in its pure form, privatizes both profits and losses. Since the decision maker "answers to" himself for the consequences of his decisions, he is responsible in the operational sense. In pure privatism the system may be said to benefit from *intrinsic responsibility*. (But, as we shall see, pure privatism is rare.)

In sharp contrast to privatism, commonism privatizes the gain but commonizes the losses. The herdsman keeps the gain (increase in his herd) for himself while pushing the losses (in the form of environmental degradation) off onto the entire community. Such a system fails to meet the operational criterion of positive responsibility.

Criticism of the commons may be made even stronger by enlarging the concept of responsibility, following an example set in mathematics more than two thousand years ago. The power of the number system was greatly increased when the human mind conceived the paradoxical idea of negative numbers. The real world presents us with no example of a negative quantity of things or substances. Negative numbers may not be "real," but they are a great aid in thinking about practical affairs. How would we manage financial transactions without the concept of debt? A debt can be thought of as a negative amount of wealth. We have been so long accustomed to this insubstantial concept that it takes imagination to picture the mental distress of the people who were first confronted with the newly invented negative numbers. The history of mathematics shows that each significant expansion of the idea of number was met with passionate resistance.

The idea of *negative responsibility* is likewise a paradoxical concept, but an immensely useful one. The unmanaged commons exhibits negative responsibility, since it actually pays the individual decision maker to make the wrong decision. It is this negative responsibility that generates the tragedy.

In socialism both gains and losses are socialized. At first glance this system might appear to be almost identical with private enterprise, with the community replacing the individual as both the actor and the acted-upon. In practice, however, the properties of this system are different, because "the community" is an abstraction, and abstractions can neither make decisions nor be rewarded.

"The state," "the nation," or "the community" cannot make decisions: only individuals can. Under socialism, individuals have to be delegated to carry out tasks for the community. That being so we must ask, are the motives of the delegates identical with the desires of the rest of the community? We dare not assume this identity. The uncertainty is basic to socialism.

The Intrinsic Defect of Delegation

The need to get a job done correctly leads the apparatus of the state to put into writing some sort of job description. However, reality can never be completely captured by words, so the job description is the beginning of trouble. Above everything else a manager wants to keep his job, so he pays very close attention to the written description though this may mean losing sight of the reality behind the words. When it becomes apparent that the description has been counterproductive, new specifications are worked up and the process of demand and response goes through another cycle. There seems to be no end to this process, particularly not in a legalistic society. What it all boils down to is this: *Meeting the*

TABLE 9-1

The three basic distribution systems, their operational meaning, and the kinds of responsibility they create

Name of system	**Operational meaning**	**Kind of responsibility**
Privatism	Privatized gains and privatized losses	Intrinsic
Commonism	Privatized gains and commonized losses	Negative
Socialism	Socialized gains and socialized losses	Contrived

requirements is easier than doing the job right. Such is the intrinsic defect of delegation.

This defect is not limited to socialist organizations. Private enterprises suffer from it too. The seriousness of the defect increases with the size of the organization and with size-related feelings of impersonality. The quantitative relation between size and seriousness is not a rigid one: the effects of overall size can be lessened by breaking the whole up into smaller operational units so that closer attention may be given to the psychological well-being of each unit. Business enterprises that compete with each other in the marketplace are under more pressure to do this than are socialist governments that have no competitors (or think they have none).

In an effort to motivate managers, socialist states establish systems of rewards and punishments for good performance or poor. With these additions the political system becomes one of *contrived responsibility*. The relations between the types of distribution systems and the kinds of responsibility are shown in Table 9-1.

Managers, in their efforts to survive, try to escape society's contrivances. They try to get control of the information system so that news of their shortcomings will not reach their masters. Some will sabotage the information-gathering apparatus if they can. This reality is deeply based on human nature, on the will to survive. Socialism is theoretically less committed to individualism than is privatism, but the individual socialist's will to survive is just as strong. The facilities for hiding blunders are undoubtedly greater under socialism.

In illustrating the limitations of socialism, rather than point the finger at Russia it will be better to look at ourselves. Though the United States is nominally a nonsocialistic country, it is, in fact, a mixed economy. Many of its subdivisions are run in a more or less socialistic way. The Defense Department is wholly socialistic. It is not governed by the ordinary laws of competition and profit and loss. It is run by managers who, like all managers, seek to control the traffic in information. Because the rhetorical shield of "national security" gives them a splendid excuse for bottling up information, cost "overruns" on weapons procurement usually have to surpass 100 percent before the public belatedly learns of the waste.

Investigation by a Senate committee in 1983 uncovered some fantastically large wastes in procurement. The committee found that there was no auditing of the purchase of spare parts when the contract was for less than $500,000. The results of this lackadaisical policy were predictable in a broad way, though the details were shocking. A hammer that cost $17 in the open market cost the Navy $435: a 12-cent Allen wrench cost the Air Force $9,606; and the Navy paid $1,075 for a bolt worth 67 cents. The prize of the lot was an antenna rotor pin with a retail value of 2.4 cents, for which the Air Force paid $7,417—a price greater than the market price by a factor of 309,042.

Such are the hazards of socialism, whether in the United States or Russia. I cite our Defense Department only to make the point that the contrived responsibility of socialism makes possible the sabotage of information systems. Only periodic investigations by other governmental units, coupled with the ever-present threat of investigations, can keep costs from escalating out of sight.

I have said little about the shortcomings of a private enterprise system. It has some. A thorough discussion of the comparative merits of privatizing and socializing gains and losses under various circumstances would make a book much larger than this. I gladly pass up the opportunity of writing it. My primary concern here is with the fatal defects of commonism, a politico-economic arrangement that is not confined to primitive herdsmen in Africa or preindustrial England. Not many assertions in political science are true when put in dogmatic form, but we can have confidence in the truth of the following statement:

> *The overall system of a sizable community struggling to survive in a crowded world may be either socialism or privatism. Either system may work, more or less. But, except in non-critical areas of distribution, commonism cannot possibly work for very long.*

The words "sizable" and "struggling to survive" in the statements above are of crucial importance, as will be explained in Chapter 12.

The Blindness of Wordsmiths

Karl Marx never mentioned Lloyd or the concept of the commons; we must presume that he knew nothing of the 1832 lectures. He was not exceptional: very few Victorians appear to have read Lloyd's work. It disappeared from the mainstream of political thinking, to be reprinted as an economics classic only in 1968. In the interval the failure to distinguish between commonism and socialism—or, with greater precision, between commonizing and socializing gains and losses—resulted in a literature that muddled economic thought at its source.

In *The Communist Manifesto* of 1848, Marx wrote: "The theory of the Communists may be summed up in the single sentence: Abolition of private property." Seventy years later the philosopher Bertrand Russell, using the word "socialism" to refer to the same system, wrote: "I think we shall come nearest to the essence of Socialism by defining it as the advocacy of communal ownership of land and capital." Both definitions focus on the relatively trivial fact of "ownership," while ignoring the critical act of distribution.

As concerns all the parts of my body—fingers, toes, et cetera—ownership may be said to be a fundamental fact. But for all the things outside my body—house, land, stock certificates, annuities, *et cetera ad infinitum*—ownership is more abstraction than fact. The point is well made in the old saying, "I don't care who owns the cow; I just want to know who gets the milk."

The principal importance of "property" lies in its effects on the distribution of income. With individual private property the effect is obvious; but when we approach community property the mode of distribution becomes the critical question. Not only Marx and Russell, but countless other influential theorists failed to keep their eyes on the ball of distribution. Result: a thoroughly confused literature.

Words can so easily mislead us. In form, these two sentences are identical:

(A) Jones owns the land; the proceeds go to Jones.
(B) The community owns the land; the proceeds go to the community.

When we pass beyond words to operations we find that the second sentence raises problems not found in the first. In (A) the two parts of the sentence stand in a one-to-one relationship. In (B), the relationship is many-to-many, and we don't know whether the two "manys" are (or should be) equal or not. Do the proceeds go to every last member of the community? In equal amounts? Is the distribution unconditional with respect to need, merit and tradition? Who, exactly, oversees the distribution? Who oversees the overseers? If there are disagreements in the community (as there certainly will be), how are they to be resolved? By majority rule? Dare we insist on unanimity? And how shall we deal with discontented minorities?

Sentence (A) defines privatism; its meaning is almost self-evident. Sentence (B) is ambiguous: we don't know whether it defines commonism or socialism. We must decide whether we want to commonize or socialize the distribution before we can figure out the effects of the system. We are unlikely to think straight so long as we use the traditional but ambiguous term "communism."

As has often been observed, no great modern state can be assigned wholly to one politico-economic distribution system or another. Socialist Russia is shifting ever more to private enterprise agriculture; "capitalistic" America is becoming ever more of a welfare state because of our compassionate desire to distribute basic necessities in accordance with need. There is little to be gained from arguing over the single adjective to be applied to an entire national state.

It is much more important that we examine each element of a society to see whether distribution is privatized, commonized, or socialized. Time spent in getting our verbs correct does more for clear thinking than splitting hairs over the meaning of the nouns.

CHAPTER 10

The Double C–Double P Game

Was there ever a poacher who was overwhelmed by a feeling of guilt when he was caught? It is doubtful. Shame, perhaps, but not guilt. The mind of the poacher lives in a past in which the abundance of resources justified the Marxist rule "to each according to his needs." (Yet the poacher may object strenuously to a welfare state, in which the same ideal justifies support of the urban poor, who, so to speak, poach on tax receipts rather than wild game.)

Does the manager of a factory feel guilty when he is charged with vastly overloading a community sewage system for which his organization is not paying appropriately high taxes? Probably not. He will likely offer the justification that his business is "creating employment" and "stimulating community development." He will see nothing inconsistent in his overdrafts on the commons of the community sewage system while he praises free enterprise and condemns the welfare state.

Self-interest and general principles are often at war. Inconsistencies are endemic in long-established societies. To say that an entire society follows the path of democracy, socialism, or commonism is to gloss over much variety in economic arrangements. It is better to examine one aspect of society after another to find the verb—the action-name—that best fits what people actually do in each case. Consequences are produced by actions, not by names. Policies must be judged by their consequences.

The Commons of Insurance

Insurance was invented long before Christ was born: the first kind may have been marine insurance. Losing an entire ship and its cargo could be ruinous to a small operator. By creating a pool of disaster money, a group of trustworthy operators could insure themselves against infrequent individual losses. When such losses are due primarily to "bad luck" (e.g., a storm at sea) rather than individual incompetence, joining in a mutual insurance society is a rational action.

Insurance commonizes losses. This is all right as long as the existence of the insurance does not encourage "disasters." Human nature being what it is, it is only prudent to see to it that the compensation paid by the insuring agency is always somewhat less than the actual loss; in this way the insured operator is motivated to do his best to avoid disasters. Less than full insurance gives the insured person "a piece of the action," making his goal the same as that of the insuring agent. This is a sound principle to follow with all kinds of insurance.

Dishonesty is always a possibility. After falsifying a manifest to describe an expensive cargo that is not, in fact, on board the ship, a dishonest shipper may scuttle his craft and claim a larger sum than ship and cargo were worth. To keep fraudulent claims from escalating, some sort of control system is needed, and that costs money. The extra cost may be assumed by the insurance association and paid for out of higher premiums; or it may be assumed in whole or in part by the larger community, which is willing to pay for detectives and policemen to discourage all forms of dishonesty. The cost of a police force is generally commonized.

Similar risks afflict all forms of insurance. At first glance, life insurance would seem exempt: the purchaser of the insurance is betting against an event (the loss of his life) that he presumably does not want to occur. Suicide, however, threatens the system. A person who has suffered what is in his mind a devastating business failure might be tempted to take out a large policy naming his wife as beneficiary, and then promptly to do away with himself. To guard against such events, insurance policies specify a period after their purchase (typically two years) during which the policy is voided by provable suicide. After that the company treats the impulse to suicide as just one more of the normal hazards of life.

Arson is more threatening to an insurance system than suicide. The successful arsonist can strike again, and skillful arson can destroy the evidence. Arson + a false inventory = a profit—at the expense of the other insurance holders, the other contributors to the common pool. In the nature of things the aggregate loss due to arson is unknowable, but it is believed to be considerable these days.

The Wedding of Commonized Costs to Privatized Profits

What is the free enterprise system? Our stereotype of it is of fair competition in the marketplace, in which enterprisers win or lose on the basis of their efficiency in running factories, effectiveness in managing labor relations, ingenuity in devising sales campaigns, and honesty in delivering quality. Some enterprisers win, some lose. Calling the system a "profit system" is misleading, because it is truly a "profit-and-loss system" as far as the competitors are concerned. The general public wins because competition ensures low prices. The great fortunes made by some enterprisers can be viewed as commissions for helping to keep prices down for everyone.

Unfortunately, the truth is not always so simple. A comprehensive history of great business fortunes would show a disconcertingly large number that were made in a quite different way: the enterpriser devised a silent way to commonize costs while continuing to privatize the profits. We will encounter this explanation repeatedly as we probe deeper into the workings of society. The system just described needs a label.

The hidden rules of the game are these: Commonize Costs and Privatize Profits. The result we may refer to as the "Double C–Double P Game," or even more briefly as the *CC–PP Game.* Such a union of privatism and commonism is not even hinted at in the official apologies for free enterprise. Those who cling to an exalted view of free enterprise should view this union as something of a morganatic marriage. Though unblessed by authority, it is rich in consequences.

Enterprisers never broadcast the information that they are playing the Double C–Double P Game: it would not be in their interest to do so. It generally takes an outside observer to be aware of the truth. An early exception to this rule may be found in the 1556 work of Georgius Agricola, who, in *De re Metallica,* gave a fair summary of the arguments against mining (from which activity Agricola himself profited). The following translation is by Herbert Hoover:

> The strongest arguments of the detractors [of mining] is that the fields are devastated by mining operations, for which reason Italians were formerly warned by law that no one should dig the earth for metals and so injure their very fertile fields, their vineyards, and their olive groves. Also, they argue that the woods and groves are cut down, for there is need of an endless amount of wood for timbers, machines, and the smelting of metals. And when the woods and groves are felled, then are exterminated the beasts and birds, very many of which furnish a pleasant and agreeable food for man. Further, when the ores are washed, the water which has been used poisons the brooks and streams, and either destroys the fish or drives them away.

Commonize costs and privatize profits—*but don't tell anyone.* This has been a formula for success for centuries. Though Agricola was not aware of it, some of the most distressing costs of mining are health costs. Until the development of nationalized schemes of compensation in the twentieth century the costs of deteriorated health were "paid" by the miner himself, partly in medical bills but even more in reduced capacity to work and enjoy life.

Premature death has characterized the vocation of mining from the earliest days. Particular causes include chronic lead poisoning, mercury poisoning, the "black lung disease" of coal miners, and poisoning by radon gas in uranium mines. The list is a long one. From prehistoric days to the present, mining enterprisers have prospered at the expense of those who did the actual work. This is true in countries like the United States, where subsurface resources are private property, as well as in countries like Mexico, where the federal government claims all wealth beneath the surface.

Variations Without Limit

Mining is not exceptional: ill health is a cost incurred by many industries. Considering that cost is a central concern of economics it is astonishing how long economists delayed in paying attention to the assignment of health costs. Gentlemen did not call attention to the Double C–Double P Game.

One is justified in suspecting that financial support has had something to do with this silence. "Whose bread I eat, his song I sing"—and the song is loaded with euphemisms that spare the feelings of mine host. Let us not forget that economics, so useful a study for business, has been generously supported in the universities by business interests.

The Nobel Prize–winning economist Milton Friedman, a dedicated supporter of free enterprise, is notable for the vigor with which he has repeatedly taken businessmen to task for keeping so quiet about business violations of the spirit of free enterprise. In 1983, in a lecture on "The Suicidal Impulse of the Business Community," he said: "With some notable exceptions, businessmen favor free enterprise in general but are opposed to it when it comes to themselves." Acting in concert they do even more harm to their cause: "The broader and more influential organizations of businessmen have acted to undermine the basic foundation of the free market system they purport to represent and defend." Commonization is the principal tool of the all-too-unconscious sappers supported by the business community.

Subsidy is one of the disguises of commonization. An airport approved of by the government will be 90 percent paid for out of federal funds. The rest is paid for by the local government and the airlines. This means that taxpayers, whether they fly or not, pay most of the capital cost of airports. Such subsidy is, in the operational sense, irresponsible. It is no surprise that so many airports, particularly in small cities, are grandiose far beyond ascertainable need. In an irresponsible system, provincial megalomania triumphs over national rationality.

A very effective way of hiding the fact of subsidy is to lend money at "concessionary" rates of interest, or to calculate the cost-benefit ratio of a proposal using unrealistically low rates. For many decades the building of great dams with federal money was justified by assuming a rate of interest that was less than 2 percent per year. With such a rate it is easy to justify many a financially unsound venture. But when federal subsidies are involved, few are so rude as to breathe the word "commonism."

The profit motive is not the only force that leads to demands for subsidy. In recent years the idealism of environmentalists has been a major force. The pollution caused by the burning of fossil fuels could be avoided by going to solar power. But the cost of capturing the energy of the sun is, at present, greater than the cost of power from coal or oil. With enough experience the comparative advantage may be reversed. Therefore, say environmentalists, let us subsidize solar power now; let's give "tax breaks" to the householder who installs a system that is, in the strict economic sense, wasteful of present resources.

Thus are environmentalists, usually so critical of businessmen, led to promote the same CC–PP Game. When an environmentalist plays this game there is a slight change in the payoff. The *commercial* profit goes not to him but to the enterprisers who sell the hardware. These agents often adopt a low profile, letting idealistic environmentalists bear the brunt of public criticism. Even environmentalists gain profit of a sort—a *psychic* profit derived from feeling that they are making Tomorrow happen. Success, if it comes, may retroactively justify the CC–PP Game; however, failure certainly won't.

"Pork-barrel" legislation is an egregious example of commonization achieved in a convoluted way. A pork-barrel bill is supported because each individual legislator sees the bill as a way of commonizing the costs of a regional extravagance that will earn him brownie points among his constituents. Benefits in this case may not be privatized in the ordinary sense; rather, they are confined to a constituency that is far less than the population that pays the bill. The congressman from Tennessee gets his dam built by promising to support the building of a dam in Arkansas, in exchange for support from the congressman from Arkansas for a dam in Tennessee. When such mutual back-scratching is general enough, a majority of the congressmen can be made parties to the commonization of a long list of dubious projects that create monetary inflation in the future.

In the aggregate, the most significant example of silent cost commonization has probably been that which hides under the name of "externalities," a term coined by Alfred Marshall, I believe. If the smoke from my factory causes a housewife to spend more on laundry soap (and spend more time at the washing machine), the cost is called an external cost of the business enterprise, or an "externality."

What a marvelous euphemism! The implied meaning is clearly this: "external to the accounting books of the firm producing the pollution." The euphemism illustrates well the principle that language is as much action as it is description. The word "externality" has the effect of pushing costs away from the speaker and toward an otherwhere too vague to evoke emotion. An external cost is a cost imposed on the public without its consent.

How one wishes Marshall had proposed the word "imposition" for these "external" costs! That might have made the public look sooner for causative agents to whom the true bills could be presented. Perhaps an even better term would have been the word "excretion." Factories excrete smoke, sulfur dioxide, and other noxious waste products, just as living bodies excrete carbon dioxide, skatole, urea, and other products useless to them. We no longer permit an individual to deposit the excretions of his body wherever he likes; neither should we permit factories promiscuously to broadcast theirs.

Owners and managers of business enterprises would not have liked such linguistic precision, of course. It is now too late in the day to replace "externality" with "imposition" or "excretion." A pity.

Taboo, the Mother of Confusion

Personal freedom, free enterprise, concessions, subsidies, externalities, bankruptcies, and shared risks—each of these terms has a legitimate use that may involve no intent to deceive. But it should be apparent by now that these terms are also admirably suited to covering up the reality of commonized costs. Personal success often depends on playing the CC–PP Game *without others becoming aware of the rules that are being followed.*

As was pointed out earlier, William Forster Lloyd introduced "commons" into the economic literature in 1833, but neither Marx and his followers, nor spokesmen for the business community, picked it up. The idea is one that revolutionists and profit-seekers would just as soon not call to the public's attention. As for the Marxists, they find that the eloquent phrase "to each according to his needs" appeals strongly to the self-interest of the needy. A dispassionate analysis of the operational properties of the act of commonizing would have no such appeal. No doubt members of both camps will continue to avoid speaking of the inherent properties of distributional systems; as far as such advocates are concerned, the word "commonizing" is best put under taboo. An effective taboo is worth more than the most skillful argument. As a countermeasure, the first step in emasculating a taboo should be to pin the proper label on it. As with Rumpelstiltskin in the fairy story, the full power of a taboo depends on its remaining unnamed.

What Can History Justify?

Before leaving the topic of externalization as a cryptic form of commonization, we need to face an awkward question: Where would civilization be today if economic externalization had been nipped in the bud at the beginning of the Industrial Revolution?

Success in business generally depends on small profits, consistently maintained. If all the costs of manufacturing and mining had been internalized in each enterprise from the very beginning, would the Industrial Revolution have been able to get off the ground? Objective data and novels from Charles Dickens to Upton Sinclair convince us that literally millions of human lives were seriously damaged and shortened by the externalities imposed on miners, chimney sweeps, lace-makers, felt-makers, steel mill workers, and laboring children. The accumulated capital that made the amenities of our world possible was taken out of the hides of common men, women, and children.

Had this injustice not been perpetrated it is possible that we might never have become rich enough to subsidize the medical research needed to prove the harm that manufacturing and mining cause. Or we might have eventually made this progress anyway, though after a much longer period. We don't know. Would the harm incurred by slow progress have been greater than the harm caused by the hell-for-leather pace actually followed? Again, we don't know.

We should be reminded of a related point raised by Marx's alter ego, Friedrich Engels, in his *Anti-Dühring:* "Without slavery there would have been no Grecian state, no Grecian art and science, and no Roman Empire . . . no modern Europe . . . no modern Socialism." Did Engels, who worked all his life for the advancement of socialism, mean to defend slavery on the ground that it produced desirable consequences? It seems most unlikely. He merely pointed out an interesting historical possibility. It seems in the highest degree probable that Engels, had he been asked, would have said that slavery, regardless of its historical consequences, must (from the present time on) be regarded as indefensible.

In the same way, and for much the same sort of reason, I think most people living today would say that even if it is historically true that the widespread externalizing of business costs was causatively responsible for the rise of modern civilization, we cannot, from here on out, tolerate the practice. Regardless of the past, future policy must insist on internalizing the true costs of production. Preferably, external pollution should be prevented by measures taken within the firm, e.g., the purchase of filtering and scrubbing equipment, the capitalized cost of which is factored into the price of the firm's product. Where that is impractical the cost of area-wide measures can be socialized (not commonized). To externalize a cost is to subsidize it. Those who vigorously oppose government subsidies should, to be consistent, also oppose the impositions of hidden costs on unconsenting individuals, costs that pass under the uninformative name of "externalities."

In passing, let us note that the situation is not significantly different in the U.S.S.R., according to the testimony of a disillusioned Russian bureaucrat writing under the pseudonym of "Boris Komarov." (See his book *The Destruction of Nature in the Soviet Union,* 1980.) The American Soviet-watcher Marshall I. Goldman confirms Komarov's findings. The resistance of Russian managers of state industries to the internalization of their costs is every bit as vigorous as the resistance of American managers of private enterprises to internalizing theirs. The pattern is the same. The human motives are the same. Only the labels differ. This is one more reason why we should pay more attention to operations and their consequences than to political labels.

It is the fashion for nations to erect memorials honoring "The Unknown Soldier," in recognition of the multitude of men in uniform whose heroism has not been specifically rewarded with individual medals and citations. Perhaps we should erect two more monuments, one to the Unknown Worker and the other to the Unknown Citizen, as symbols of the millions whose unrequited suffering from "externalities" over the centuries made possible the industrial momentum that propelled Western civilization to so high a material level in so short a time. Without their sacrifices, the life of the average citizen today might not be significantly better than it was at the time of Christopher Columbus.

CHAPTER 11

Compassion vs. Principle

"To steal from the A&P is fine, but to steal from a little grocery store run by an old couple is unthinkable." The speaker was a young radical in Boston, and the time was June 1970. Middle-aged readers of *Time* (where this was reported) may have been shocked by the denial that stealing is always wrong, but the reported conversation truly reflects a common attitude (though many sharing this belief might be unwilling to state it so openly).

There is an essential conflict between compassion and principle. To give the most favorable interpretation to the words of the Boston radical, let us first suppose that he is speaking of stealing from a chain store to get food for a poor neighbor. In such case we must grant that he is moved by compassion. The verifiable pain attributable to the loss suffered by any stockholder of the store would be minuscule compared to the verifiable pleasure brought to the recipient of the stolen groceries. Such might be the radical's argument. Obviously this argument makes no allowance for the tragedy of the commons; it does not recognize the bearing that time and the changes brought about by time have on moral problems.

But when the radical steals from a single old couple who owns a Ma-and-Pa grocery, the harm done is done in the present and is undiluted by large numbers. Recognizing this harm requires so little imagination that even an utterly present-bound mind can comprehend it. It is only when we take time seriously that we recognize that compassion must have limits.

A Utilitarian View of Compassion

In the survey of the effects of externalities, pork-barrel legislation, and subsidies, we have seen that time-blind self-interest promotes commonism. This should be no surprise: every widespread practice is promoted by self-interest, often (as with commonism) in disguised forms. But commonizing would not be nearly so widespread if it depended only on egocentric motives. *Homo sapiens* is a social animal who is both selfish and caring of others. Purely as a motive, compassion for others is beyond criticism. But consequences are at least as important as motives in the judgment of moral actions. Compassionate actions that increase loss and suffering cannot be recommended as public policy.

The conflict of altruistic and egocentric impulses creates ambivalence. How much sharing should one indulge in? How much can one afford? Given a number of alternative actions, how does one choose among them? Economics, an ego-based discipline, has a simple answer to its particular problem: Always try to maximize the economic return to Ego. But what is it that one tries to maximize through altruistic actions? The number of recipients? The total income of all recipients? The warm feeling that Ego gets from "doing good?" The well-being of the next generation? Human dignity? . . . There are too many possibilities for comfort.

Philosophers have proposed a number of ways of measuring good. Without judging their comparative merits I will here adopt the utilitarian perspective, because it seems closest to what the man in the street—and the woman as well—is guided by most of the time. "Utility" is a measure of the happiness or pleasure that an action brings *to the totality of those affected by the action.* Utilitarians want to maximize utility. Obviously pleasure is a difficult thing to measure, but in some approximate way utilitarians try to sum up the various negative and positive utilities of an action before deciding whether to recommend it or not.

Utilitarianism cannot be charged with being a selfish system of ethics, since it considers the consequences for all the participants. Problems arise when we try to agree on the rules for reconciling the conflicting values of the various actors in the moral drama.

Let's look closely at what happens when our young radical steals from a store. As far as the reckoning of materials is concerned, stealing is a "Zero-Sum Game." What one party to the transaction gains, the other party loses. For this part of the reckoning it doesn't matter whether the losing party is a chain of supermarkets with an aggregate income of $100 million a year, or a retired couple fleshing out a pension with the modest income from a corner grocery store.

The utilitarian calculus deals not with material wealth but with feelings—feelings of pleasure or pain. The units of utilitarianism are psychological units. What feelings of loss or gain in happiness are created by the two alternative thefts? The stinted couple running the Ma-and-Pa grocery may feel keenly even the tiniest of thefts. By contrast, the corporation that owns a chain of stores can hardly be said to feel anything. As for its thousands of stockholders, it would be surprising if the most energetic shoplifter could diminish their dividends by a penny. The direct psychological cost to the shareholders resulting from shoplifting by a single individual is essentially zero. By the simplest utilitarian reckoning, individual thievery from a chain of stores can be treated as a Non-Zero-Sum Game.

Now let's look at the situation through the eyes of the radical thief. This time we will suppose he is stealing "for his own account." He can justify his action by comparing these two composite utilities created by the theft:

(A) plus-utility to me + negative-utility to Ma-and-Pa
(B) plus-utility to me + zero-utility to the supermarket

In the first case the radical thief is willing to admit that the negative utility may be far greater than the positive; being compassionate, he refuses to indulge in such thievery. In the second, however, he says that the aggregate utility is positive. The radical's happiness appears to cost the corporation nothing—so why not go ahead and steal? The thief thereby increases the total amount of happiness in the world; isn't that a good thing?

If such thievery were the only behavior justified by this rhetoric, we might just label the argument sophistical and pass on to other matters. But (as we shall see) the same kind of argument is tacitly employed by respectable citizens to justify their actions, so we need to explore the logic further.

The fault of the utilitarian argument as developed so far is that it is not ecolate. It does not ask the question "And then what?" If for the moment we grant the validity of the thief's argument, what happens when others follow his example? The negative-happiness caused by one theft may not be perceptible to the stockholders of a large corporation, but the consequences of ten million thefts will. Unpunished thievery is "catching." The ultimate consequence of permitting acts that are individually barely harmful is the creation of a commons in which people steal from stores in proportion to their needs (as they are pleased to perceive them). The result is a disaster for all of society.

To be acceptable, utilitarian ethics must be ecolate, must take account of time and the unfolding of human behavior as actions become precedents. This conclusion is not new: Immanuel Kant expressed the same idea two centuries ago in his Categorical Imperative: "Act only according to that maxim by which you can at the same time will that it should become a universal law."

The mature theory of utilitarianism makes a distinction between act-utilitarianism and rule-utilitarianism. Our Boston radical was an act-utilitarian. John Stuart Mill and most of his followers are rule-utilitarians: with scarcely a glance at Kant, their calculus deals with the consequences of rules rather than of mere acts. The ecolate question "And then what?" leads to the path Kant and Mill blazed.

Right-wingers who belabor utilitarianism, moral relativism,

pragmatism, and rationalism sometimes have a point. When a rational approach to moral problems is so shallow as not to include time and the consequences of time—the effects of precedents—then indeed can a rational approach that seems benign in the short run turn malign in the long. Compassionate jurors who vote for an overgenerous award in an injury case may bask in a feeling of warm benignity, but as their decision becomes a precedent, a multitude of similar judgments sharply raises the cost of insurance to everyone.

One of the curious features of the behavior labeled "conservative" in our time is the way it combines time-sensitive approaches to conventional moral matters with time-sensitive views of economic behavior. A conservative of this stripe condemns induced abortion because it "eats into the moral basis of society," while he calls for a higher birthrate because it "is good for business." He sees the effects of a high birth rate as this cascade of consequences:

more babies————→more customers
more customers————→more business
more business————→more profits
more profits————→more prosperity for all

What is not figured into this calculus are other consequences of the second in the series: more customers———→ greater demands on the environment, causing an earlier exhaustion of nonrenewable resources and a higher production cost for other resources, thus decreasing real income per capita.

Rules and their verifiable consequences are a major concern of most thoughtful utilitarians, who might justly be called ecolate utilitarians. The ecolate question "And then what?" brings time into the ethical calculation. Since time has no stop, but ethical cogitation must have, it is a matter of judgment how far the cascade of possible consequences should be pursued before settling on a decision. Ecolate utilitarians always run the risk of becoming indecisive Hamlets. However, society at large is threatened less by those who try to see too far than it is by those who want to stop an inquiry after going only as far as the step that yields them a personal profit.

Bankruptcy, a Challenge to Compassion

That a man should be held responsible for his debts seems only just, yet adhering to rigid accountability sometimes produces results that seem neither compassionate nor wise. In many societies in the past a man who was unable to pay his debts lost his civil rights and sometimes had to wear special dress in public. In Dickens' novel *Little Dorrit* the father was confined to debtors' prison, from which position he could neither discharge his debt nor pay for his upkeep. True, the imprisoned debtor's situation might serve as an instructive example to others who might be tempted to follow his imprudent path, but was not the cost to society rather large? This consideration led principled citizens to lean toward a more compassionate conclusion. In industrially advanced countries debtors' prisons are now a thing of the past. (In poor countries, debt-peonage—really a form of slavery—is still common, and almost as difficult to escape from as a debtors' prison.)

Bankruptcy laws permit a bankrupt to discharge his debts, i.e., not to pay them and not to be saddled with them as he tries to make a new life. You can call this unjust (to the creditors) or compassionate (to the debtor), as you wish, but it does help harness a man's abilities to society's work. It is not uncommon for a commercially adventurous man of ability to go into bankruptcy several times before he makes a fortune. It is probably beyond our ability to design an empirical study that would show the net effect of bankruptcy laws on the well-being of society as a whole.

The immediate effect of bankruptcy on creditors is hardly good (save that it does educate them to be more cautious in the future). The essential effect of legal bankruptcy is to create a commons limited to the bankrupt's creditors who, in retrospect, can be said to have maintained a pool of resources from which their late customer took "according to his needs."

With a change in the scale of bankruptcy new consequences appear. When a company as large as Chrysler is in danger of going "belly-up," new strategies of survival become possible. When it became obvious that this large automobile company was in imminent danger of bankruptcy some of the more principled supporters of the free market insisted that nature be allowed to take its course, to emphasize that what we carelessly call the "profit system" is really a profit-and-loss system. Forgiving a loser today, it was felt, would run the risk of escalating bankruptcies in the future. The suspicion was a reasonable one, but nonetheless Chrysler managed to wangle a bail-out from the government. To date, the salvage has not led to a rash of imitators.

Had this large corporation gone belly-up, countless of its suppliers would have followed suit. Some banks might have had to call on the resources of the Federal Deposit Insurance Corporation to avoid closing their doors, and this would have created a drain on the federal treasury (which is part of the national commons). Unemployment would have pyramided, calling for more federal funds. Acute awareness of what was happening might have established a vicious circle of fear-engendered retrenchments in the business world. The very scale of the threatened failure called for a sort of preemptive compassion not available to a small firm.

In the event, the government made the needed funds available by purchasing stock warrants that would have been worth nothing had the rescue ultimately failed. Since these purchases were made from public funds, this means that Chrysler, by virtue of its size, was enabled to dip into a nationwide commons. . . . The ways of a nation officially committed to "private enterprise" are sometimes marvelous to behold!

By the principles of private banking practice the Chrysler rescue effort was unwise. But in fact, in the end, the government more than got its money back. One wonders: will this success story make it easier for large companies in the future to escape the much-praised discipline of the marketplace?

The Medical Commons

The function of socialized medicine is strictly Marxist: it assumes that meeting medical needs is an obligation of the entire community. In Great Britain the National Health Service performs this function.

In the United States we have a more complex system. Many private health plans (e.g., Blue Cross) commonize the costs among their subscribers. The costs of a few spectacular diseases (e.g., end-stage renal disease) have been commonized over the whole nation by Act of Congress. (If you need a kidney transplant Uncle Sam will pay for it.) And postretirement medical costs are paid for in large part by federally funded Medicare. The system is in a state of flux, but the trend toward commonizing medical costs is clear.

So what? No one *wants* to be sick, so surely there is no danger that the commonized costs will escalate? Alas, the situation is not quite that simple.

First of all, personal choice does enter in. Perhaps no one wants to be sick, but many people want to live the sort of life that favors illness. A recent study showed that 13 percent of hospital patients accounted for 50 percent of medical costs, and that smokers and drinkers were more common in the more expensive portion of the population.

Americans have the right to drink and smoke and do many other things that are injurious to their health. If all medical bills were paid by the individual, there would be no reason to question that right. But to a large extent the bills for unwise living are paid for by everyone. Does the right to live unwisely include the right to impose extra costs on those who are wiser and more abstemious?

One physician answered that question in the negative in the pages of the *New England Journal of Medicine* in 1971; his article provoked the greatest number of letters, both pro and con, in the long tenure of the editor. Physicians, many of whom smoke and drink, have as much trouble as the rest of us in reconciling freedom with responsibility, compassion with accountability.

Medical bills in industrially advanced nations amount to about 10 percent of the Gross National Product. Socializing medicine diminishes forces that tend to contain costs. Because of our laws, the upward pressures work in a unique way in the United States. Though quite unplanned, there is cooperation between the medical and legal professions in bringing about the escalation of medical costs.

In the first place, doctors have little motivation to hold down costs. On the contrary, doctors who own a share of a private community hospital can quite unconsciously opt for a patient's hospitalization in case of doubt. The conflict of interest is obvious but seldom mentioned in public.

The most blatant extravagance takes place with heroic medicine: prolonging the lives of those certain to die soon, transplanting organs, and saving the lives of "preemies" (prematurely born infants) whose vital natural functions have to be propped up with advanced and expensive technology. Organ transplants cost from $100,000 upward, a preemie below two pounds is likely to cost a quarter of a million dollars, and keeping aged and unconscious human vegetables alive for years sometimes runs into the millions of dollars.

To what end?

Defensive replies to the last question tend to be wholly literate, wholly verbal: "the sanctity of life," "the unconditional obligation to save lives," and the like. Adding economic considerations to medical problems is regarded with horror by those who are satisfied with mere eloquence. Yet the reconciliation of conflicting goals demands that principles be made numerate as well as literate.

The smaller a preemie is, the greater the cost of saving it. Since small size does not stop us from trying to save preemies, an external observer of the earthly scene—the traditional

Martian—would have to deduce that we believe that the smaller and less mature a life is, the greater is its value. If a two-pound baby is worth a quarter of a million dollars, what is the value of a one-pound baby (the like of which we have not yet saved)? A million dollars? Ten million? . . . How about a one-ounce embryo? Saving such a one will no doubt soon be technically possible.

Ecologic/economic studies of any of the other ten million species of plants and animals show incontrovertibly that, by any rational standards, the value of immature individuals is least for the earliest stages of development. We note that the instincts of animals leads them to act as if they understood utilitarian ethics. Puny or helpless babies are sooner or later abandoned. Animal parents that invested heavily in grossly immature offspring that were almost sure to die would leave fewer breeding offspring in the next generation than would animal parents that followed sound bioeconomic rules. Natural selection favors bioeconomic rationalism.

Of course, human beings are free to practice as aberrant behavior as they please. The question is, how long can the species survive practicing unnatural ethics? Technology escalates the cost of heroic medicine at a rate far exceeding the rate of increase of the national income. Will the day arrive when we put all our income into saving nearly invisible embryos, patching up the nearly unpatchable, and preserving aged human vegetables, thus leaving no income for the enjoyment of life by those who must work to pay the bills?

Actually, the present situation has not arisen out of reflective planning but merely from thoughtlessly following certain historical lines of development without taking note of the consequences. For a physician to try his hardest to save a life did little harm so long as medicine was relatively impotent. Heroic medicine is a child of modern technology which has grown explosively in the past half-century, giving physicians powers Hippocrates never dreamed of.

The media have made a bad situation worse by reporting the latest medical extravagances as if they were sporting events.

What records have been broken today? The smallest preemie saved? The most difficult organ transplanted? The most miserable life forced to continue?

Three great classes of interests are ignored in the usual journalistic treatment. First there are the interests of the individuals "saved." The later life of ex-preemies is often a nightmare of chronic medical troubles; however, long before this stage has been reached, the reporters have disappeared.

Second there are the interests of the family. Even when most of the economic costs of heroic medicine are picked up by a foundation or by the nation, the life of the family is cruelly distorted by the insistent personal needs of the flawed human being. The siblings of a defective child often suffer greater psychological damage than does the defective child itself.

Finally, there are the interests of the people at large, among whom the economic costs are commonized. Commonizing dilutes the intensity of the felt cost, but the public purse does have a bottom. The fantastic sums spent on caring for spectacular, newsworthy cases probably makes it harder to allocate money for less spectacular but more widespread illnesses, particularly of the poor. It may also be true that some of the most promising and ambitious young doctors move into the kinds of practice that capture the biggest headlines, though their abilities are more needed elsewhere.

The legal profession has unwittingly worsened the situation. The minority of doctors who are incompetent open up broad horizons for lawyers specializing in malpractice suits. In the United States (though not in England) such suits are handled on a contingency fee basis. Typically the law firm collects nothing if it loses the case, but 30 to 50 percent of the award if it wins. Such a gamble motivates the lawyer to give the case his best try—and some very sharp lawyers are attracted to the field.

To counteract this threat, a physician takes out malpractice insurance. The lawyers hired by an insurance company are paid for out of the premiums imposed on the doctor, who recoups his cost through higher bills to all of his patients.

Knowledge of the existence of malpractice insurance is not admissible in the courtroom because jurors are supposed to reach their verdict without considering the financial resources of the defendant. The exclusion is wise, but it doesn't work. For decades jurors have known that virtually all physicians are insured, so they assume that which the rules of evidence are designed to keep them from knowing.

When the jury retires to consider the evidence, compassion wrestles with principle. The jury knows that the pleasure the plaintiff will get from a large award—even after losing half of it to his lawyer—is far greater than the individual displeasure such an award will impose on any one of the millions of people over whom this commonized cost is spread. Therefore . . .

The steps in the escalation of medical costs need to be spelled out. High awards by juries cause malpractice insurers to raise the premiums demanded of physicians, who of course have to increase their charges to patients. This increases the cost of medical insurance for the patients. Malpractice insurers put pressure on doctors not to overlook any possibility in examining the patient. Where one or two laboratory tests used to suffice, a doctor may now demand twenty. This is called "practicing defensive medicine." (It is more defensive for the doctor and the insurer than it is for the patient.) The premiums for medical insurance rise still further.

The higher the cost of personal medical insurance, the greater grows the demand for federal subsidy, for completely socialized medicine. Even those who realize that it is the commonizing of costs that is producing the runaway escalation find it difficult to suggest a politically acceptable way of stopping the ruinous process.

What began as a call for compassion in helping the sick has now ended up in a demand for a "right to medical care" for all. The invention of new "rights" is one of the major indoor sports of our time. Most of the new sportsmen are literate, non-numerate, non-ecolate folk with only a tenuous grasp of time and the power of runaway feedback.

The dream of limitless medical care for all is a dream of a complete medical commons. This will be an unmanaged commons if the right to care is unqualified, or if the law continues to allow medical practice to be harassed by aggressive lawyers and soft-hearted juries who do not understand the ultimate consequences of their decisions. The tragic end can be prevented only by introducing *No!* into the system at some point or points. What should we limit? The kinds of defects and malfunctions eligible for treatment? The ages at which heroic medicine may be practiced? The situations producing liability? The size of awards to plaintiffs? The amount of the fees charged by lawyers? We might begin by following England's example and forbidding lawyers to take malpractice cases on a contingency basis. England's experience indicates that depriving the attorney of the fortune-or-nothing gamble results in his devoting only a reasonable amount of time to the case.

Every point in our medical and legal system seems to be biased in favor of compassion. At some point compassion must yield to principle.

CHAPTER 12

The Effects of Scale on Values

From Plato's time to the present, professional philosophers have too often tried to solve problems of "the good" without considering how potentialities, behavior and value are affected by scale. Science has taken a different path. As far back as the seventeenth century, Galileo gave sound mathematical reasons why a mouse simply cannot be as big as an elephant. The weight of an animal goes up as the cube of its linear dimensions, whereas the strength of its supporting limbs goes up only as the square. From this simple mathematical difference profound practical conclusions follow.

Simplifying Galileo's treatment, suppose we compare two identically shaped animals. Animal A is 3 units long (never mind what the units are), while animal B is 6 units long. How do their weights compare?

Weight of A = 3 cubed = $3 \times 3 \times 3 = 27$
Weight of B = 6 cubed = $6 \times 6 \times 6 = 216$

We can see that 216 is 8 times as great as 27; though animal B is only 2 times as long as animal A, it is 8 times as heavy. (Note that 2 cubed is 8.)

As for the strengths of their legs:

Strength in A = 3 squared = $3 \times 3 = 9$
Strength in B = 6 squared = $6 \times 6 = 36$

So B's legs can bear only 4 times as much weight as A's legs. But B is 8 times as heavy (and has the same number of legs), so B's legs are only half as strong as they need to be (4 divided by 8). Clearly something has to change as the length of an animal increases.

If the material of which the legs are composed is the same, then the cross-sectional area of the leg has to be doubled. The leg has to be thicker.

This is not all. Since heavier bones will require more powerful muscles, many secondary changes in appearance will also be necessary, not to mention the behavior which must correlate with the lightness or massiveness of the body. If mice evolved to be as big as elephants, their silhouette would be that of elephants. The daintiness of the mouse would be replaced by the ponderousness of the elephant. If evolution produced such a new kind of a mouse, we would be inclined to view it as some kind of an elephant. Thus does simple mathematics prove the point that a mouse cannot be as big as an elephant.

In reaching this conclusion we have to do a bit of calculating the first time; but after that we need only remember the general pattern of the calculation. We need, one might say, to program into our brains the general idea of the "scale effect." In the first instance we prove the thesis by a numerate argument, but thereafter the name alone is enough to recall the numerate underpinnings of the conclusion.

Much of the practice of science requires measurements of great precision; but there are also many areas in which only the crudest of measurements suffice to establish important conclusions. The direction of research is often decided by back-of-the-envelope calculations for which precision is not required. The practice of numeracy includes not only precise figuring of the sort demanded of bookkeepers, but also the rough-and-ready estimating that often suffices to get the big picture. The scale effect is a prime example.

The scale of things determines what is functionally best. This applies not only to the shapes of animals but also to the shapes of airplanes and countless other human artifacts. Scale is also relevant to the problems political and moral philosophers deal with, though the formal theories so far developed in these fields seldom deal explicitly with this effect.

Perhaps no shortcoming of utopian thinkers is as striking as their apparent blindness to scale and its implications. A politico-economic system that works well with small numbers may fail utterly with large. This is one of the most important factors accounting for the ultimate failure of utopian communities.

The success or failure of communities based on privatism or socialism is hard to predict, but of a commonistic community it can be firmly asserted that it is sure to fail—unless it remains small enough to enjoy certain advantages of scale. This was hinted at in the statement made in Chapter 9 by words that will now be italicized: "For a *sizable* community struggling to survive in a *crowded* world . . . commonism cannot possibly work." Both of the emphasized words imply quantities. We will take up the second, first.

Scale Effect and the Cowboy Mentality

In an uncrowded world like the one our ancestors enjoyed in pioneer America, a commonistic arrangement may be satisfactory. Fruit taken from common-property trees present in excess, or game animals harvested from vast wild herds, do not demonstrably diminish the resources available next year. Commonizing a cost of zero hurts no one. Similarly, wastes may be thrown away into vast areas without harming other people, so long as the metabolic powers of uncrowded nature are more than sufficient to recycle the elements.

What Kenneth Boulding calls the "cowboy economy" is a rational and acceptable adaptation to frontier conditions. In such a world it would be silly and wasteful either to try to fence in the game as a form of private property, or to pay public guardians to watch after the common property.

But the politico-economic system that works well on the frontier breaks down miserably in a world as crowded as ours. Unfortunately, long after the reality has vanished, the dream of an uncrowded world endures, often romantically glorified. The persistence of this sentimental dream is one of the things that accounts for the clear conscience of the game poacher: to him the commons seems somehow the morally superior distribution system.

The dream also accounts for some bizarre proposals of the less insightful of the "libertarians," a group that would like to simplify economics by doing away with all community supervision. In the name of individual freedom, both poachers and libertarians would unwittingly recreate the commons with its tragic consequences. But the words "commons" and "commonizing" are missing from their vocabularies.

In a trivially abstract sense, would-be modern cowboys may have a good idea, but the scale is wrong. The judgment of "good" must be tied to scale.

Scale and the Effectiveness of Shame

"Society in every state is a blessing," said Thomas Paine in his treatise *Common Sense,* written in the year of the Declaration of Independence. He then went on to say: "Government, even in its best state, is but a necessary evil." What he, and many following him, did not see was that the relative importance of the two words "necessary" and "evil" changes as a function of the size of society. The larger the community, the more necessary is some form of government. A large population with no government will soon disintegrate into the evil of chaos.

Paine at least admitted that government is necessary. Many utopian thinkers have denied this. I think it is no coincidence that the greatest efflorescence of utopian thinking and utopian experimentation coincided with the most rapid population growth of the Western world. The same factor that made the growth of government inevitable caused restrictions to increase at so fast a pace as to be obvious to all and onerous to many. The two hundred years after Paine have been the heyday of utopian experimentation.

Hundreds of communities based on commonized distribution systems were established in the United States in the nineteenth century. There was a second flowering in the 1960s. Most of the groups disintegrated within a few years. In all this plethora there are only a few success stories, of which the story of the Hutterites is the clearest example. Why their utopia has survived more than three generations—the minimum standard of meaningful survival for a social experiment—tells us much about commonism.

These people, united by religion, live in agricultural colonies in the northwestern United States and adjacent areas of Canada. Followers of Jakob Hutter (burned as a heretic in 1536), they live by Marx's ability/needs equation. They are immensely successful. They accept the best of modern agriculture and medicine while rejecting all forms of birth control, even the rhythm method and celibacy. As a result they are the fastest-growing population in the history of the world, with the average number of living children per Hutterite woman being close to ten. At the present time the total number of American Hutterites is on the order of 50,000. (Non-Hutterites may be pardoned if they wonder about tomorrow—and the day after tomorrow.)

Why have thousands of other idealistic communities committed to the Marxist ethic failed to persist? "From each according to his ability, to each according to his needs" sounds simple enough, but it evades this question: Who decides what a person needs and what he can do? If the state decides, that's socialism. If each individual decides for himself, that's anarchism or commonism. When dreamers set up idealistic communities they almost invariably opt for commonism, since it appears to be most compatible with the selfish and individualistic impulses that have been selected for over thousands of millennia, the selection being operative long before *Homo sapiens* evolved. The chosen system often works for a while, but then it breaks down. Why?

The Marxist system is critically sensitive to scale. Hutterites found this out long ago when they noticed that as their community grew in size, the relative numbers of goldbricks increased. Workers found excuses for going to town "to get a part for the tractor" or whatnot, lingering for a long time before returning. What were at first the actions of a few became the actions of many, as some of the hardest workers decided they were unwilling to carry the burden of the whole community. Altruism diminished as envy and resentment took over. Instead of working, people argued about working. Exercised abilities declined; expressed needs increased.

The answer, the Hutterites found, lay in controlling the scale. As long as the community was less than some apparently critical number (about 100 to 150), the Marxist distribution system worked. Above that not precisely defined number the system failed, and failed ever more badly as the number increased. So the Hutterites adopted a development program that plans for the automatic splitting of a community into two as soon as its numbers have doubled. One farming community becomes two, two become four, and so on, at intervals of about fifteen years. So long as abilities and needs are determined within a really small community, commonism works.

Why the sensitivity to scale? What is the force that keeps individuals from abusing the right to determine their own abilities and needs in a small community, but not in a large? The effective force is *shame*. (This word is seldom heard in our society these days, a fact worth pondering.) A potential goldbrick in a small community can be shamed by public opinion into doing his share of the work and not taking more than a fair share of the proceeds. As the community grows in size, the effectiveness of shame diminishes. Why? Perhaps because each individual is less visible. Perhaps because goldbricks offer each other mutual support. Perhaps . . . but this is a subject for further research.

The essential point is that commonism will not work in a

community that is "sizable," where this word means more than 100 to 150 members. The statement of this great principle of political economy is only roughly numerate, but it suffices to rule out of consideration most utopian writings, which show no consciousness of the effects of scale.

No purely commonistic community that transcends the Hutterian limit has persisted for three generations. Mormonism, for instance, which started out on a commonistic basis, survived only because it soon changed to a combination of muscular private enterprise and closely monitored religious socialism (to take care of its charitable welfare activities).

Numerical limits are observed in many of the elements of societies that may, or may not, be commonistic overall. For instance, China is now trying to persuade families to keep the number of their children to one or none. In part, coercive laws are employed, but raw coercion seems to be less important than shame.

Shame, which is sensitive to scale, can have little effect in a nation of over a billion people acting as a whole. Nationwide laws are not enough to bring about the needed control of population. Fortunately for the new population policy, the Chinese nation is broken up into "production groups," which seem (so far as we can judge from the outside) to fall below the Hutterian limit. Within these groups, in response to national limits imposed on each production group, women actively use shame to control each others' reproductive behavior.

The Chinese also have found that limited production groups are a necessity for the maximum production of many goods, particularly of farm products. According to a 1983 article in the government-owned *China Daily,* the earlier commonistic systems have now been replaced by what the Chinese call a "responsibility system," in which each small production group is allowed to keep, or sell in the open market if it wishes, any production that surpasses the goal set by the national government. Thus, says a government journalist, China has established a "new socialist principle," namely, "From each according to his abilities, to each according to his work."

Thus has the wheel been reinvented once more. Russia also, faced with the devastating nonproductivity of socialized agriculture, is painfully moving toward privatized production of farm products, in which farm workers are rewarded for work, not needs.

The Importance of Management Costs

The loosely quantitative adjective "uncrowded" in our basic position statement must be interpreted in terms of the exact segment of the distribution problem to which we apply it. For instance, a city of several million people can hardly be said to be uncrowded, and yet we allow access to its sidewalks on a commonistic basis. Taxes that pay for the construction and maintenance of the sidewalks are paid for by everyone according to their (presumed) ability to pay, but the sidewalks are available to all on the basis of need, as determined by each individual. The result is not disastrous. Why not?

At the risk of laboring the obvious, we spell out the details of the successful commonization of the sidewalks. The principles involved have general application. A commonized sidewalk is not likely to be overused because each person is physically limited in the amount of "need" he can muster. If someone wants to walk twice as fast and thus "use up" the sidewalk at twice the average rate, who cares? So small an excess exacts no perceived sacrifice on the part of those who demand less. A very fat man may require more sidewalk space; but again, who cares?

Theoretically we could charge individuals in proportion to their quantified use of the sidewalk, but it is obvious that the management costs of so doing would be quite out of proportion to the good that could be achieved. So all of us—economic libertarians, capitalists, socialists, everyone—cheerfully take care of our sidewalk needs on a commonistic basis. An ideologue who refused to do so would be judged rather silly.

Even so, this decision is implicitly based on scale. At some high level of crowding we might decide that we needed another system, probably a socialistic one. We have already had to reach this painful decision with respect to streets-as-parking-space. Unmanaged, on a first-come-first-served basis, insufficient spaces favor those whose time is worth relatively less and who can afford to come early to get a parking space. Those who have more money than time resent the result and press for another solution.

Parking meters proctored by municipal employees restrict access on a socialistic basis. Alternatively, privately owned parking garages deal with the problem within the framework of privatism. Which of the two systems (socialism or privatism) is superior in this instance is a matter on which intelligent people can legitimately disagree. But once the scale of cities has passed a certain point, if time is of any value, there is no rational defense for the commonistic arrangement.

In choosing one system or another we must always keep in mind the danger of escalation. If my neighbor is the hospitable sort who fills his house with guests every weekend, it would be foolish of me to worry about the extra load on the commons of the sewage system. The management costs of making the charges "fair," i.e., strictly proportional to use, just aren't worth the possible gain.

But sewers serve factories as well as homes. The new class of customers calls for a change in the rules. Industrial wastes need to be treated differently than domestic wastes; and industries differ. Unfortunately, after the philosophy of a commonized sewage system had been accepted for many decades people forgot (if they ever knew) that the argument for the commonized system assumed the standard and nonescalating "needs" of domestic households. The argument took no account of the far from standard "needs" of industries, needs which are all too capable of escalating, with disastrous consequences for communities that fail to make the necessary distinction.

It is not cost-effective to put a pollution meter on the

sewage outflow of every household, but something of this sort is needed for every industry that asks to be hooked up to a municipal sewage system. Otherwise, a businessman who praises the free enterprise system can benefit from a system that he would no doubt condemn in other contexts, namely commonism. Playing the Double C–Double P Game, he undermines the economy of the community that has invited him in.

PART II

THE LARGER VIEW

CHAPTER 13

The Global Pothole Problem

Once upon a time there was a city whose streets suffered a pestilence of potholes. Plainly more tax money was needed to fix the streets, but the people marched to the slogan of "No unfair taxes!" Since every new tax is unfair to somebody, the mayor could not find the money needed to fill the potholes. Things went from bad to worse, until finally the holes became so monstrous that they broke the springs of the mayor's limousine and His Honor had henceforth to come to City Hall in a three-ton truck. The mayor decided things had gone far enough. He asked the local Genius for advice.

"The answer is simple," said the Genius (after spending six months and seven hundred thousand dollars on a study). "I have made a survey of all the nations and have found potholes everywhere. Clearly we are confronted with a Global Pothole Problem. Everything is connected to everything else. Global problems call for global solutions. If we want to get our potholes filled, we must establish a Global Pothole Authority.

"The GPA will be responsible for first surveying and studying the pothole problem, following which it will resurvey and restudy it. At some point in time it will undertake to fill in the potholes. For uniformity and fairness, all requisitions for this work, from whatever part of the world, must be processed by the central office of the GPA in Geneva. Approval will be based on need. Financing will be by taxes based on national ability to pay. This means that for many years to come all of the potholes filled will be in the poor countries, while the taxes will be levied only against the rich. This is only fair.

"Let me emphasize to Your Honor that this is a great opportunity for polishing up your image as a citizen of the world. By taking the larger view, the global view, you can strike a blow against parochialism, provincialism, bigotry, and selfishness. Global thinking is the mark of the truly civilized man. Under your inspired leadership, our city can make the future happen."

The regional Council of Churches and the local chapter of the United Nations Association got behind the proposal, and the Global Pothole Authority was born. The future began to happen.

Unfortunately the city's potholes remained unfilled. The Genius took his fee and bought a cottage in a fashionable lakeside community; he wasn't going to let the potholes bother him. The mayor continued to ride to City Hall in a truck.

Why "Global"?

In 1933 the great 12-volume *Oxford English Dictionary* identified the word "global" as *rare*. From the examples given, it would appear that the word was at that time no more than a superfluous synonym for "spherical." During the following half-century "global" acquired new meaning and a halo. The establishment of the United Nations in 1945 marked a change in the climate of opinion. "Global" became an incantation whereby a speaker could signal to others that he was a person of unlimited compassion. The word acquired a fashionable aura.

Aura is nice, but what about the word's operational meaning? Operationally, the word "global" indicates a desire on the part of the speaker to commonize regional expenses over the whole globe. Since the regions that would benefit most are usually poor ones (and not the speaker's), the motivation is apparently unselfish. It is casually presumed that every ubiquitous problem is a "global" problem. But "ubiquitous" is merely a descriptive word, whereas "global" has now become a prescriptive term, a term that implicitly prescribes political action, generally the commonization of costs.

Death, taxes, poverty, and sin are ubiquitous, but do we want to define them as global, in the new sense? Is it advisable to try to commonize the costs of these ancient ills?

To speak of "global poverty" is to imply that commonizing wealth will put an end to poverty. "Global hunger" implies that commonizing food resources will put an end to malnutrition and starvation. To speak of the "Global Population Problem" is to imply that an overpopulated nation cannot control the breeding of its own people. But if a unitary nation cannot control nature's reproductive force, what chance is there for a consortium of bickering nations? Or for people living under the anarchy of no nations?

Globalistic thinking was greatly encouraged in 1968 by the introduction of the seductive concept of the "global village," an invention of Marshall McLuhan, a Canadian professor of communications. What the paradoxical combination implies is simple and indefensible, namely that there are no scale effects in political affairs.

Whether justified or not, our nostalgic view of small villages is that they are places where life is simpler and hence sweeter. McLuhan connected this vision with communication: in a small village everybody can talk with everybody else. By the grace of modern technology it is, in a certain theoretical sense, now possible for everybody in the world to communicate with all others, though there are now nearly five billion human beings in the world. But—a *village* of *five billion* people, all talking?

Misunderstandings are surely some sort of power function of the number of people seeking to be understood. Information suffers from diseconomies of scale. Communication technology is a mixed blessing. There is no way that a true village can be created out of large numbers. Numeracy repudiates the purely literate dream of the global village.

Now that computers are in the ascendant, we are inclined to look to them for solving problems dealing with large numbers and many groups. But detection of the folly of globalizing problems requires no more sophisticated a computer than the unaided human brain. Ancient wisdom is all the software we need.

Long experience has shown that local problems are best dealt with by local action. Various scale effects account for this fact. Increase in numbers brings an increase in the possibilities of misunderstanding, an increase in the necessity for delegating, and an increase in the ways that delegation can malfunction. Other things being equal, large agencies are less efficient than small. The reason for this is quite simple. Self-interest urges individuals to evade responsibility whenever they possibly can. The more distant the monitor, the more feasible evasion becomes. Globalization favors evasion. The wise rule to follow should be plain:

Never globalize a problem if it can possibly be dealt with locally.

All this is so obvious that it should not need saying. However, the rise of environmentalism has been accompanied by a great increase in the volume of voices calling for global attacks on all problems, great or small, extensive or localized.

Globalism is usually counterproductive. This is not to say that there are absolutely no problems that are truly global. There are: not many, but a few. And they pose difficulties far more serious than most globalists realize. In serious matters fashion is always a peril. The next chapter deals with a potential problem that will, if it develops, prove to be truly—and staggeringly—global. Before we clasp globalism to our collective bosom, we had better understand the implications of a truly global problem.

CHAPTER 14

A Truly Global Problem

It doesn't pay to go back on your promise—not if you are dealing with a god. Cassandra, daughter of Hecuba and Priam (the last king of Troy), promised to shack up with Apollo if he would first bestow on her the gift of prophecy. He did so. Then the foolish girl defaulted on her promise. Why, one wonders? "Apollo" is the very name for male beauty; moreover, he was a god. Perhaps he lacked a subtle something.

Be that as it may, the divine one dealt with the perfidious maiden in a devilishly clever way. He left her in full possession of her prophetic ability but decreed that henceforth no one would believe her.

The Meaning of Cassandra

So goes the Greek myth. Like most primitive wisdom it is presented as a story about persons, a common form for humanity's first attempts at understanding human behavior; the scientific focus on the mechanisms of behavior comes later. Even at this late date a personalized story is more effective in capturing people's attention than is a pure description of mechanisms. For that reason, let us bend the truth just a bit as we describe a little-known visit of Cassandra to the President of the United States.

The frustrated seer was trying to enlist the President's aid in alerting the public to the dangers of overheating the earth through the "greenhouse effect" created by the discharge of ever greater quantities of carbon dioxide into the atmosphere.

THE PRESIDENT: How long did you say it would be before we are in serious trouble?
CASSANDRA: There's no certain answer. Maybe 50 years; maybe a century or more.
THE PRESIDENT: In less than four years I am up for reelection. You surely don't expect me to put my political future on the line for a disaster that may only possibly occur long after I'm dead?
CASSANDRA: But think of posterity!
THE PRESIDENT: Sorry, I've got to think first of myself, of the next election four years from now. Between you, me and the gatepost, I will admit that the substantive content of my public statements doesn't matter a hell of a lot. But the color, the flavor of what I say matters a great deal. Surely you ancient Greeks knew that "Honey catches more flies than vinegar"? I intend to catch a lot of flies, a lot of votes, in the next few years. I'm not about to scare the bejesus out of my supporters.

The mechanism of standing for reelection in a representative democracy tends to suppress unpleasant prophecies. The more distant the prophecy, the less a candidate can gain by voicing an unwelcome truth.

The same mechanism operates in private businesses, as we have seen in the reaction of American automobile executives to the premonitory signs that smaller cars were coming. The difference between business executives and elected officials is not as great as many detractors of bureaucracy would have us believe. Both are subject to all-too-human pressures not to disturb our equanimity by looking beyond a rather near horizon.

In group dynamics short vision usually pays off better than long. To lose in the struggle for the group's attention is to lose power. History is written by the winners. Education in the broad sense—which includes all forms of advertising—is hemmed in by the necessity to please.

All parts of the economic system conspire to deprive Cassandra of power. There's nothing personal in this conspiracy. We must remember that the word "conspire" literally means "to breathe together." Impersonal mechanisms cause the participants to breathe together, as it were, to keep Cassandra from being heard. When we understand how the world works we realize that there is no Cassandra-*person,* only *Cassandraic mechanisms* that insure that precious few people will believe and act upon unwelcome predictions.

Belshazzar's Feast

Though we are reluctant to admit into consciousness any predictions about the future that are unpleasant *for us,* we are captivated by accounts of others who have been forced to face their Cassandras. The account of Belshazzar's Feast in the fifth chapter of *Daniel* fascinated Rembrandt and other painters. An imaginative person shivers at the vision of King Belshazzar recoiling in horror from the disembodied hand writing on the wall those mysterious words, *Mene, Mene, Tekel, Upharsin.* We have no trouble believing that "the king's face changed color, the joints of his loins relaxed, and his knees knocked against each other."

Fate can be delicious—when we see it being imposed on someone else. We love it when Daniel says that *Mene* means "God has numbered your kingdom and brought it to an end," while *Tekel* means "You have been weighed in the balance and found wanting."

Curiously, the Biblical tale has Belshazzar reward Daniel for the frightening translation. It would appear that the Hebrews of the second century B.C. had not as sophisticated an understanding of human nature as that achieved earlier by the Greeks who gave us the story of Cassandra.

Carbon Dioxide = Mene, Mene?

The disembodied hand of Belshazzar's Feast has now become the instrumented hand of impersonal science and technology. Unfortunately it writes too many things on the wall. For the sake of our sanity, we have learned to ignore most of the messages. But one of the messages, updated month by month and year by year, is beginning to seep through to consciousness. This is the record of the carbon dioxide content of the air, as measured in the winds that sweep past the four-thousand-meter peak of Mauna Loa in Hawaii. (See Figure 14-1.)

Those who put the most pessimistic interpretation on the findings of science worry that this curve may be our *Mene, Mene, Tekel.* Our Daniel is the generic Scientist: what does he say? To get to the (possible) bad news first, we wonder if this jagged line may not be a record of the first tiny steps on the way to Earth becoming another Venus, a planet whose surface is 611 Fahrenheit degrees above the temperature of boiling water. As climatologist Stephen Schneider put it: "The ever-rising curve from Mauna Loa has taken on the characteristics of a cobra poised to strike."

Comparing Venus to Earth we note that Venus's mass is 82 percent as great, its gravity 91 percent as great, and its distance from the sun 72 percent as great. The relative amount of solar irradiation per unit area is 192 percent that of Earth. Because radiative loss of heat follows a fourth power law, "other things being equal," the greater amount of radiant energy received should make the temperature of Venus only a little bit higher than that of Earth. It was long supposed that these facts made Venus a sort of sister planet to our earth, one on which humanity might someday live.

Then space probes showed beyond doubt that the mean surface temperature of Venus is 730 degrees Kelvin, as compared to 280 degrees Kelvin for Earth. That translates as 823 degrees Fahrenheit on Venus compared to an average of 44 degrees Fahrenheit on Earth. No one had expected so great a difference. Moreover, the atmospheric pressure on Venus is 95 times that on Earth, and the Venusian atmosphere is rich in sulfuric acid. Venus is hardly a place we would even want to visit, much less inhabit.

Why is Venus so hot? More to the point: why is Earth so

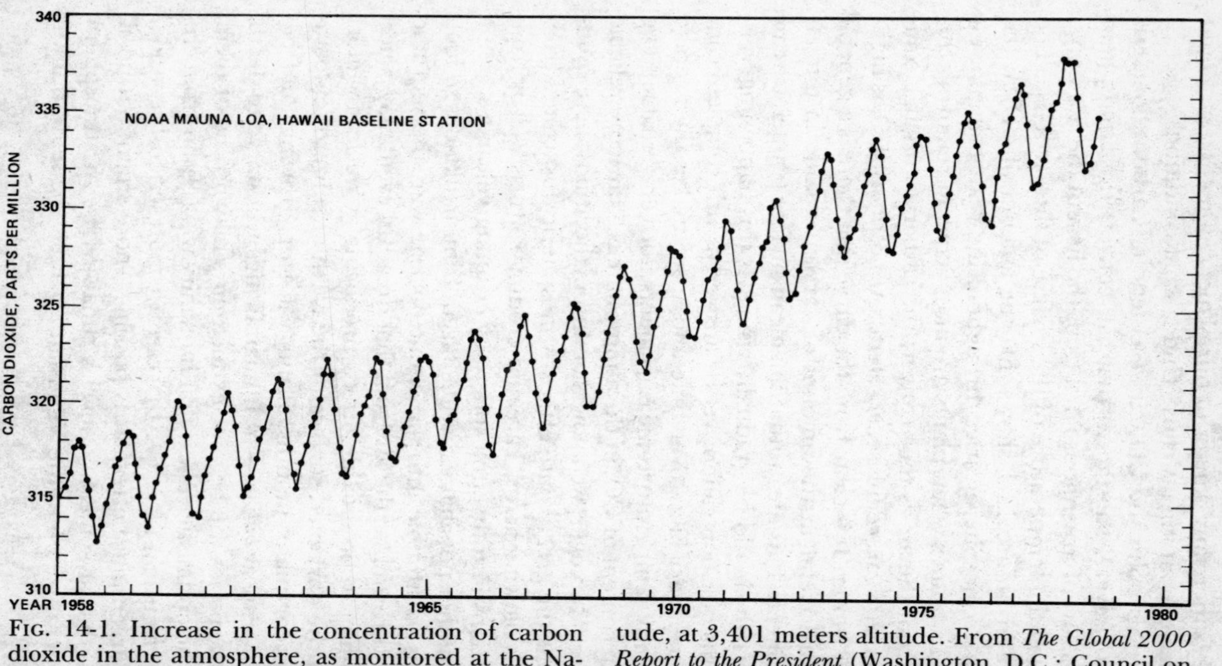

FIG. 14-1. Increase in the concentration of carbon dioxide in the atmosphere, as monitored at the National Oceanic and Atmospheric Administration's Mauna Loa Observatory, Hawaii, 19.5 degrees N latitude, at 3,401 meters altitude. From *The Global 2000 Report to the President* (Washington, D.C.: Council on Environmental Quality, 1981).

cool? Is there any danger of our planet getting hotter? Could we become another Venus? Under what conditions?

At this point the CO_2 curve enters in. Carbon dioxide is one of several substances that can convert the atmosphere into a sort of greenhouse that traps the heat of the sun's rays, making the temperature of the surface of the earth rise. The curve in Figure 14-1 shows that the amount of CO_2 in the earth's atmosphere varies throughout the year, being lowest in the summer when plants are most active in removing this gas from the air as they convert the carbon to organic compounds (sugar, cellulose, et cetera). Atmospheric CO_2 is highest in winter. It is clear from the curve that there is a general upward trend in the amount of carbon dioxide in our atmosphere. This means that the mean surface temperature of the earth should be gradually rising. Is this happening?

This question cannot yet be answered with complete confidence. Variability over time and space makes the average of many temperature measurements not very reliable. The mean content of CO_2 in the atmosphere is a far more reliable figure. Moreover, we know that dust in the atmosphere counteracts the greenhouse effect. Atmospheric dust comes from winds blowing across bare and eroding soils, as well as from injections into the atmosphere by erupting volcanoes.

Another possibility must be faced. If humankind is so foolish as to stumble into an all-out nuclear war, the soot and dust thrown into the air from that dreadful event may cancel out the carbon dioxide effect, producing a "nuclear winter" with the darkened surface of the earth at subzero temperatures for many months. Plants will stop growing, and billions—repeat *billions*—of human beings will starve to death. Many species will become extinct; ours may be one of them.

The optimistic view is that humanity will refrain from creating a nuclear winter. In that case, we must seriously consider the greenhouse effect. It is possible—not certain, but possible—that at some CO_2 level a runaway rise in atmospheric temperature up to the Venusian level may be set in train. It is believed that such a runaway process must have occurred on Venus at some time in the past.

Don't go out tomorrow and try to buy tickets on an interstellar spaceship headed for other stars, other planets. Disaster may not overtake us. Some providential and natural corrective feedback may save us. Human intelligence may save us. Maybe so, maybe not. We don't know. We simply do not have enough facts—yet. But the mere possibility that Figure 14-1 may be our *Mene, Mene, Tekel* is reason enough to keep our eyes open and look for more facts.

In the meantime, we can learn more of the current "informed opinion" about the greenhouse effect while we await the gradual replacement of informed opinion by accepted facts. In what follows I rely principally on a Conservation Foundation report prepared by Walter Orr Roberts, for many years director of the National Center for Atmospheric Research.

Contributors to the Greenhouse Effect

Inasmuch as carbon is subject to natural cycling, the upward trend seen in Figure 14-1 is a bit of a puzzle. Extra carbon dioxide results in slightly faster plant growth, which removes CO_2 from the atmosphere faster. This negative feedback ("corrective feedback") should produce a flat trend line, but it doesn't. Moreover, plants are not the only things that remove carbon dioxide from the air. The ocean can absorb this compound, making it part of a carbonate-bicarbonate complex. Evidently it can't absorb carbon dioxide as fast as we inject it into the atmosphere.

The overload is accounted for by the extra burning of coal and oil by industries serving an ever larger population of human beings. The carbon in these fuels was captured from the air millions of years ago by photosynthetic plants and then locked away underground until man came along and released the "fossil sunlight." The carbon we now set free in a single year took some 400,000 years to capture and store. The extra quantity of carbon dioxide is evidently too much for nature's adjustive system to take care of. Presumably if we stopped burning coal and oil for a decade the carbon dioxide content of the atmosphere would go down.

We are unlikely to try the experiment, for it would play havoc with our whole industrial system and with the civilization that depends on it. In fact, at present, there is even an acceleration in our rate of use of fossil fuels. Ironically, keen concern about the dangers of nuclear energy has produced political pressures to use fossil fuels at an even greater rate, thus augmenting the greenhouse effect and its consequences.

Industry puts into the air other compounds that contribute to the greenhouse effect. Nitrogen oxides from automobile exhausts and Freon (fluorocarbon compounds) from the little spray-cans that dispense shaving soap, cosmetics, insecticides, et cetera, have much the same effect. (Responding to the call for responsible citizenship, the manufacturers of many of these products have now substituted less harmful gases for Freon.)

Even methane, a common product of metabolism, adds to the greenhouse effect. Recognition of this has led to the interesting discovery that termites produce large amounts of methane, perhaps between 15 and 40 percent of the methane in the atmosphere. The quantity of metabolizing termite flesh is astonishing: it has been estimated that for every human being on earth there are about three-quarters of a ton of termites. The biomass of termites is more than ten times that of the human species.

Consequences of the Greenhouse Effect

The quantitative uncertainties affecting the causes of the greenhouse effect also bear on its consequences. More CO_2 in the air should mean a higher temperature—but how much higher? And what will be the human consequences of a higher air temperature?

At the present rate of change, the carbon dioxide content of the atmosphere might double sometime in the twenty-first century. There is a consensus among climatologists that this would produce a worldwide warming somewhere between 1.5 and 4.5 degrees Celsius (3 to 8 degrees Fahrenheit). There is also a consensus that this apparently small change would have great consequences—though exactly how great, and what kinds of consequences, are matters on which informed opinions differ. What are the possibilities?

If the glaciers and South Polar icecap were to melt completely, putting their water into the oceans, the ocean level would rise about 200 feet. This would have a catastrophic effect on many of the major cities of the world, which are down near the water's edge: London, Rotterdam, New York, San Francisco, Singapore, Calcutta . . . the list is long.

But we can put that danger out of our minds for the present. Melting the miles-deep ice of the South Pole would take perhaps a thousand years—longer than the physical artifacts of modern cities are likely to last anyway. A thousand years gives time enough for a city to be moved to higher ground. Considering the immediate problems posed by the threat of nuclear war, it would seem rather silly to worry about a danger a thousand years away.

Long before the glaciers and the South Polar icecap melted, the sea ice covering the North Pole and surrounding both poles would melt, for it is only a few feet thick. Once the atmospheric carbon dioxide content was doubled, it might take only another ten years to melt the sea ice. The direct effect of this melting would be trivial: Archimedes' Principle tells us that there would be no change in sea level. But the change in *albedo*—reflectance—in going from glittering white ice to blue water would greatly increase the absorption of heat over those areas of the globe, thus speeding up the warming of the earth. This would give us less time to adjust to the consequences of the greenhouse effect (whatever they may be).

Climatology is far from a precise science, but climatologists are agreed that even a few degrees' warming of the earth would have fantastic effects on the distribution of weather over the earth. Some deserts would become humid, while some presently humid areas would suffer from droughts. The frequency and distribution of thunderstorms, hailstorms, cyclones, and hurricanes would change. Precisely how they would change is beyond our present ability to predict. But climatologists would bet their bottom dollar that the changes would have great repercussions on economic and political affairs.

A suggestion of the sort of thing that might happen can be seen in the map shown in Figure 14-2. This shows what the weather was like around 5000 B.C., during the warmest period since the last Ice Age. The total rainfall over the face of the earth might have been no different from now, but the distribution was much different, and distribution is all important in human affairs.

Suppose that the warm period that is to come should produce a climatic pattern like that of this past period, and that the change took place in a few decades. At the present time the United States and Canada produce 75 percent of the grain exported to other countries. The major cereals grow in regions where the rainfall is only slightly greater than the minimum needed, but the precipitation is reasonably reliable. Given a climatic regime like that of 5000 B.C., it is certain that Canada and the U.S. would cease to be grain-exporting nations and would become dependent on other nations for their food. But there are no other nations that can produce enough grain to supply us. Russia, already in difficulties with her agriculture, would be in worse trouble.

Who would be better off? One looks first to some of the wetter areas on the map. But most of these are in the tropics, and their climate would be even hotter than it is now, as well as wetter. Such conditions are not favorable to cereal production. Rice would have to be the principal exportable crop

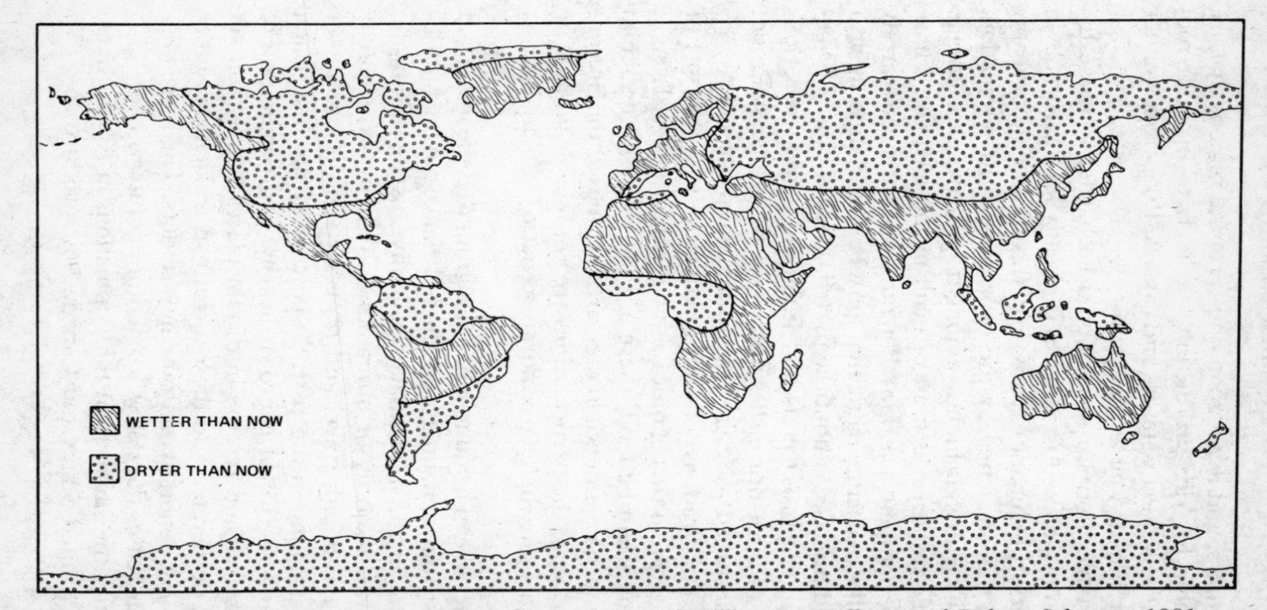

FIG. 14-2. A plausible map of the world's climate as it may become because of the greenhouse effect; based on the world as it was in the warmer Altithermal Period (4,500 to 8,000 years ago). Derived from the work of William W. Kellogg and Robert Schware, 1981, as reported in the *Conservation Foundation Newsletter* for April 1983. (Courtesy of the Conservation Foundation)

of the tropics, and it is a good one. Tropical regions that are not suitable for rice production are likely to produce root crops, such as sweet potatoes and cassava. Root crops neither keep nor ship as well as the grains.

Exportable food surpluses would be harder to come by. The ratio of people to resources is already least favorable in the tropics; the climatic expansion of this area would probably be more than matched by the expansion of its population, for fertility is already highest among tropical peoples. Storage of organic resources is most difficult in those climes. Those who have not spent a year in a hot, humid country can scarcely imagine the perpetual battle that must be waged there against fungi and insects. Without the partial die-back caused by periodic winters, insect pests and fungi are extremely serious enemies of agriculture. Pesticides and fungicides help, but their widespread use raises environmental problems.

It would be rather pointless to continue with this sort of speculative *pre*construction of the Greenhouse Age. Even at the purely scientific level, there are too many uncertainties to have much confidence in the particularities of the preconstruction. Since the particularities are not reliable it might be argued that the whole Cassandraic scenario is a mere fantasy. Not so.

For the sake of argument one can grant that the losses and the gains from climatic change might balance out on a worldwide basis. But an individual does not live his life worldwide; for the vast majority of humankind life is overwhelmingly local. Even though travel and transportation is easier now than at any time in the past, 90 percent of the food eaten throughout the world is grown nearby, and people do not pull up stakes and resettle in distant places unless their situation is desperate. And those already in residence in other areas do not generally welcome new settlers. The more population increases the less welcome immigrants will be.

Consider the situation of the populations in North America. The United States and Canada now produce about 75 percent of the world's grain exports. Suppose the weather shown in Figure 14-2 becomes a reality, and the citizens of Canada and the United States can no longer feed themselves from their own farms. Who will feed them? Argentina? Australia? South Africa? These are all possibilities.

Or perhaps needy North Americans will decide they should emigrate to some other countries. Who would take in their scores of millions? Mexico? Africa? India? Saudi Arabia? These, too, are possibilities.

What would actually happen is unforeseeable. It would depend on many unpredictable political events in the interval before our crops failed. It would also depend on the ideological positions of the more fortunate countries. Considering the large populations already living in most of these areas, how likely is it that any of the possible recipients of immigration would erect a replica of our Statue of Liberty with its motto, "Give me your tired, your poor, your huddled masses . . . the wretched refuse of your teeming shore . . ."?

Only one country in the world has such a statue, and there is no sign of any other erecting a similar one. The survival of the human species over several million years has depended on territorial behavior. There is no reason to think the future will be any different. Under conditions of shortage, failure to adopt a territorial attitude amounts to the establishment of a commons. Territoriality, in preserving the quality of life for some, increases the misery of others. Commonism brings misery to all.

Although little reliability attaches to particular predictions of the human consequences of the greenhouse effect, we can reliably predict that the interaction of physical changes with such political realities as ego-centeredness, group affiliation, and territoriality will produce a Cassandraic catastrophe. Bluntly summarized:

Any significant change in the world's weather is almost sure to be bad for civilization.

Outrageous Political Problems

Of course, humanity would adjust to climatic changes. We have done so in the past. The human costs of adjustment are closely connected with rates—the rate at which climatic change takes place, and the rate at which the necessary political adjustments can be made. Perhaps one should also mention the rate at which we learn to trust our Cassandras (but that seems to be nearly zero).

There would be little trouble if humanity were so sparsely distributed over the earth that the most rational distribution system was one in which resources were freely available in a commons. Such was probably the situation in 10,000 B.C. In a sparsely populated world, surrounded by a bountiful nature, the Zero-Sum Game played by man and nature is one in which *man* gets the pluses, *nature* the minuses (many of which she then corrects).

When crowding compels man to replace the unmanaged commons by private property or a managed commons, Zero-Sum Games then become ones in which some *people* get the pluses while the corresponding minuses devolve upon other *people*. Unfairness that may be tolerated in a long traditional Zero-Sum Game is likely to be powerfully resented in a game newly imposed on the players. A climatic shift that changes the payoffs to the players will almost certainly lead to serious political conflicts.

Population has grown past the point at which the distribution of the commons is defensible. If the greenhouse effect leads to climatic changes that significantly alter the distribution of rainfall and temperature, how will we adjust? Two possibilities must be considered. First, we may then be as we are now, a world fragmented into many sovereign nations. Alternatively, we may have achieved the idealists' dream of One World, a single sovereignty. Without asking how probable this achievement is, let us explore the consequences of climatic change in a world governed by a single sovereignty.

Indicating areas by their present national names, suppose the climatic change caused the *area* of the present United States to be afflicted with a population that was too large by 100 million people (assuming an unchanged standard of living). Two questions arise: What rational adjustments might be made? Who would call for them?

As for the adjustments, these are the principal possibilities:

- Do nothing: let individuals in that area fight it out until they have reduced their numbers by 100 million.
- Work out a program of attrition whereby the birth rate is greatly lowered until the usual death rate reduces the population to the newly required size; the extra population might in the interim be supported by philanthropic gifts from the rest of the world. (Question: How probable is that?)
- Ship the excess 100 million to the newly favored parts of the world. (Questions: Is out-migration voluntary or involuntary? Is in-migration into the receiving areas by the consent of the recipients or not? Are the immigrants allowed to reproduce at any rate they like after entering the new land?)
- Reduce the standard of living so that the entire population can continue to live where they are. (Is the reduction carried out in a market economy? If there are already economic inequalities, are these increased or lessened?)

Enough of the possible adjustments; now for our second set of questions. Who determines the adjustments to be made? A single benevolent World Dictator? Do his subjects perceive him to be benevolent? As time goes on, does he continue to be benevolent? If so, do all his successors continue in the path he has blazed?

But perhaps you don't like the idea of a global dictatorship. Perhaps you dream of the present independent nations coming together to form a single sovereign state, as the thirteen colonies came together in 1787 to form the United States of America. A plausible dream, but it needs to be made numerate.

Most thoughtful students of American history view the fusion and birth of the *United* States as something of a miracle. How much more of a miracle it would be to bring together 13 times as many sovereign units, with 2,000 times as many people speaking at least 200 times as many languages! This is to say nothing of the many diverse cultures that would have to learn to live together, or of the many religions, some of which have been warring for centuries. Think of northern Ireland; think of India; think of Sri Lanka; think of Central Africa. The dream of One World is far more daunting than getting the lion to lie down with the lamb. It is expecting several hundred would-be lions to lie down together.

By the time the accumulation of carbon dioxide demands the creation of at least a partial sovereignty to deal with genuinely global problems the population of the world will likely be some seven billion people. Assuming that the global decisions are made by a majority, one shudders to think of the subsequent reactions of the minorities.

Obviously we could go on and on with the possibilities of adjustment in One World. But let's not beat that horse to death. Let's see what might happen in a world more like ours, with more than 150 sovereign nations, a number that is increasing every decade. Suppose these numerous sovereign entities were organized into some sort of federation. (A federation didn't work very well with the American colonies, but such an organization might be the best that the international political processes can produce.) This thought-experiment produces many possible answers.

- Seeing disaster ahead, a previously rich country might start a forcible preemptive redistribution of the world's lands—the sort of thing Nazi Germany tried to do with its cry for *Lebensraum.* This word translates as "room for living," meaning room for *our* living. The consequences of such a program in a world grown nuclear are too monstrous to contemplate.
- Perhaps by this time the burgeoning idea of "rights" would have been so extended throughout the world that a newly short-changed nation would build its case on the "inalienable right" of every nation to a quality of life at least equal to the

best it enjoyed in the past. The right of one nation to have implies the obligation of other nations to supply. Do those on whom the obligation falls agree?

- A newly stinted nation, if it was fully committed to the idea of the sanctity of every nation's property (not just its own), would tighten its belt and soldier on without bullying or begging other nations. The belt tightening could take the form of lower living standards or a reduction in population. And, no doubt, a higher death rate—at least for a while.

But why go on? So many unknowns, so many unknowables, render every scenario suspect. For outrageous challenges like this there are no ready-made solutions.

Taking Cassandra Seriously

It may be tragic that Cassandra's predictions are so often rejected, but it is hard to see how human behavior could be significantly different. It would be a mistake to condemn the rejections as irrational. Most of life's decisions are made on the implicit assumption that the future will be much like the past. This has been a winning strategy for eons of time. Our species has no doubt been selected for being conservative in this sense. We are not exceptional. Assuming that the future will be like the past is the basis of all learning in all species, and most of the time the assumption works pretty well.

How can people believe in a prediction of an event that has never occurred before? The acceleration of science and technology ensures that more and more often Cassandra's predictions will be of the kind that raises this hard question. Our record in answering the hard question is not good. The threat of a serious greenhouse effect may not yet be well established, but already the evidence is good for the deleterious effects of "acid rain"—precipitation made acid principally by combustion-released sulfur compounds. The productivity of millions of acres of forests is being significantly reduced, and fish are disappearing from many streams.

Independent scientists are convinced that the acid rain danger is real, but politicians who must decide between better forests and fishing grounds or unhindered industrial growth understandably ask, What is the *bona fides* of the person who (in effect) says, "I am Cassandra"? All experts are, and should be, suspect. Even a majority of experts does not automatically command credence. Therefore (say the politicians), "Let's have another study first."

Today's response to the evidence for acid rain effects foreshadows tomorrow's response to the greenhouse effect, if the danger becomes undeniable.

If the worst-case scenario of the greenhouse effect is true—but we do not yet know that it is—it calls for an incredibly massive disruption of our lives. The burning of all fuels would have to be greatly curtailed throughout the world, which would mean lower factory output and an increase in unemployment. Muscle power would increasingly substitute for fuel-derived power (a reversal in the historical trend of centuries). Homes, offices, and factories would be colder. Warm clothes would have to substitute for much space heating. There would be less travel for all. And negative population growth might become a matter of state policy, as it is now in China.

Such is the bouquet of policy responses to the greenhouse effect that no shrewd politician would touch with a ten-foot pole.

Localization the Path to Solution

The time is not ripe for a political solution to the global carbon dioxide problem. The reason for exploring this problem in so much detail was just this: to demonstrate the difficulty of solving any problem that is inescapably global. This is the justification of the advice given in the previous chapter: "Never globalize a problem if it can possibly be dealt with locally." Once we have internalized this rule, we can accept the wisdom of the following conclusions:

- Though potholes are ubiquitous there is no "global pothole problem."
- Though hunger is ubiquitous there is no "global hunger problem."
- Though overpopulation is ubiquitous there is no "global population problem."

All three ills—potholes, hunger, and overpopulation—are produced by local action, the common feature of which is expecting or demanding too much of local resources. The mitigation of these ills requires local action; difficult as it may sometimes be, it is more feasible and more reliable than a global approach.

The puzzle is this: how did globalizing local problems become so popular? To this we now turn our attention.

CHAPTER 15

Origins of the Global Dream

Those who make science or mathematics their career must learn not only facts and theories but also methods of solving problems. Those who make political affairs their business must acquire another sort of ability: how to *not* solve problems. When a dispute shows signs of approaching resolution, a participant who fears being on the losing side shows great ingenuity in getting the issue off the track. A motion is offered to refer the issue back to committee, or to table the issue. Such a motion has the appearance of preserving the question for more study with a view to eventual solution, but the appearance is often a fiction. "Refer back to committee," like as not, means to kill the measure; but it sounds nicer. This world of contentious human beings needs a large repertoire of techniques for not solving issues (while giving the appearance of trying to do so).

Is that the explanation of the popularity of globalizing social problems? Since globalizing a problem puts it virtually beyond the reach of practical solution, do those who fear that a real solution might adversely affect their personal interests espouse globalization because they know there's no danger of its succeeding? The possibility should not be overlooked.

But such an answer may be too cynical. The global approach has a wider attraction, derived from ideas inherited from the past, particularly from our religious past. Those who never go to church may be as firmly (though unconsciously) bound by these ideas as the most ardent churchgoers.

The Global Thread in Christianity

Stereotype has it that Christianity is a global religion. The adjective "catholic," claimed by the Greek Orthodox, the Roman, and the Episcopalian churches, means simply "universal." The claim that this is the original position of Christianity is most frequently supported by the third chapter of Paul's Epistle to the Galatians. As given in the Goodspeed Bible, verse 28 reads:

> There is no room for "Jew" and "Greek"; there is no room for "slave" and "freeman"; there is no room for "male" and "female"; for in union with Christ Jesus you are all one.

The meaning of the terminal qualifying clause is made clearer by the two verses that come just before the above quotation:

> For in Christ Jesus you are all sons of God through your faith. For all of you who have been baptized into union with Christ have clothed yourselves with Christ.

What about those who have *not* been baptized? Ah, that was another matter. In our time many Christians like to think that their religion is not much given to discrimination, but this is a latter-day development. At its inception Christianity, like almost all other religions, discriminated strongly between in-group and out-group. Without this discrimination the sect might never have survived and grown. The in-group was not, however, closed, but was open to all who would "take the pledge."

The universality of Christianity lay not in its equal concern for members and nonmembers, but in its openness to new membership. In their exclusiveness Judaism and Christianity are no different from the general run of religions. Few religious leaders have been so foolish as to urge a promiscuous—that is to say, a nondiscriminating—concern for all mankind. Even when there were fewer than 200 million people on the earth, as seems to have been the case in Christ's time, doctrinal promiscuity was a luxury no religion could afford. Now that the population of the world is twenty-five times as great, discrimination is even more necessary for a religion that wants to survive.

The seeds of global thinking were present in Christianity from the earliest times. Ironically, these global seeds grew as the religion itself contracted as a living force. Better transportation and communication caused an increase in commercial relations among tribes and nations. This improvement tended to weaken discrimination, because it is hard to maintain smooth commercial relations with a trader you despise. Christianity became less discriminating and in some respects less meaningful emotionally. Following this development, the few Biblical passages that had a catholic flavor came to be taken more seriously than they were when they were first written.

Globality as an Evolutionary Stage

In the nineteenth century the global dream was joined to the idea of evolution. The new partner was not evolution in the technical biological sense, but the more general idea of evolution as incorporated in such phrases as "the evolution of mankind from savagery to civilization." Discovering new evolutionary series was a great indoor sport in Darwin's century. While Darwin stuck to his knitting, the historian W. E. H. Lecky discovered an evolutionary justification for universal nondiscrimination among human beings. In 1869 Lecky wrote:

> At one time the benevolent affections embrace merely the family, soon the circle expanding includes first a class, then a nation, then a coalition of nations, then all humanity.

Just exactly what this implies depends on what operational meaning one gives to the words, "the benevolent affections embrace . . ." Some people in the twentieth century have taken them to point to One World, a single sovereign political entity for dealing with all the world's problems. The evolutionary shift of power from family to class to nation to all humanity is seen as a decree of fate. This view gives a meaning to the word "evolution" not sanctioned by biology. The changes biologists deal with are changes brought about by selection, not by preordained decrees. The idea of preordination raises the awkward question, "Preordained by whom?" Science cannot deal with such a question.

In the rational realm Lecky's series needs to be justified by plausible selective pressures. As it happens, justification is possible for every stage in the series—*except the last*. We may in truth be under desperate necessity to reach the last stage, to establish a global authority empowered to deal with global problems; but need does not create fact. There are overwhelming reasons to doubt that the last stage in Lecky's series will ever be realized. The reasons for this assertion are based on the broadest possible meaning of "selection," a meaning broader than the biological.

Each of the various groups larger than the family arose because it gave a selective advantage to those joining such a group in competition with people who refused to become joiners. A tribe can muster more political strength than a family, and a nation can muster more political and military strength than a tribe. The advantage in each case depends on competition at the same level, with the same class of aggregations. When a class of aggregation has only a single member, there can be no competition to confer advantage to this unique group. Then competition within the group becomes the dominant factor. As competition between subgroups and individuals takes over, the unique large group disintegrates.

As Bertrand Russell said in *Authority and the Individual*: "A world state, if it were firmly established, would have no enemies to fear, and would therefore be in danger of breaking down through lack of cohesive force." The conclusion can be stated more strongly: in time, intragroup competition would be certain to shatter a single all-inclusive group.

External enemies are, in Russell's terms, the "cohesive force" that holds a nation together. National leaders have known this "in their bones" from time immemorial. When internal cohesion weakens, the shamelessly Machiavellian leader evokes or invents an external enemy. Hitler understood this. George Orwell understood it. More recently, the military establishment of Argentina showed that they understood it when they tried to strengthen their weakened domestic position by attempting to annex the Falkland Islands, which they maintained had been unjustly taken from them by Great Britain a century and a half earlier. Unfortunately for their dream of creating internal unity by inventing an external enemy, they underestimated the puissance of the external force. The literate part of Machiavelli they understood; they failed the numerate tests.

"One World," one state, if achieved would lack a cohesive force to endure. It is interesting to note that many writers of science fiction have realized this. They have postulated malevolent enemies from outer space to hold One World together. Unfortunately, in probing the only portions of space near enough to affect us in the foreseeable future, we have found no evidence of the needed malevolent aliens. Lacking the blessing of a global enemy we have yet to find the cohesive force needed for an enduring global state.

Yet the problems created by such naturally commonized evils as global air pollution call for global solutions. Lacking an all-encompassing global sovereignty, can we create a limited sovereignty, that is, a sovereignty limited to the problem it was designed to deal with? Can we, for instance, create a Global Air Pollution Board that can successfully control the amount of pollutants each member state puts into everyone's air?

The problems of subdividing sovereignty are great, but let us hope they are not insuperable. What we most fear is that the limited global sovereignty created to deal with one problem might escalate into a Leviathan that would control more than we bargained for. With greater exactness: what we fear most is that a limited sovereignty might be captured by one of the member states and converted to its own national purposes.

The fear is rational, the solution unknown. We must continue to live with the need and the fear. We must continue to look for creative solutions.

CHAPTER 16

The Paradoxical Wealth of Information

Antoine-Laurent Lavoisier, a French nobleman, was born in the year 1743. A chemist, he believed that the world was ruled by conservation principles. He lost his life to the French Revolution.

Marie-Jean-Antoine-Nicolas Caritat de Condorcet, another French nobleman, was also born in 1743. A mathematician, he believed that the world is governed by nonconservation principles. He too lost his life to the French Revolution.

Apparently there are some questions a political revolution cannot settle. The public argument over the application of conservation principles continues to this day. It is not a matter of dispute among scientists. Whether conservation prevails or not depends on the category of existence being dealt with. Those who do not understand the nature of the fundamental categories of existence continue to stir up time-consuming and fruitless controversies.

Categorical confusion leads to newspaper stories like the following fictional one, the equivalent of which seems to be printed every year:

> Scientists at Hogwash University have found that the value of an adult human being is $8.74, based on current market quotations for sodium chloride, magnesium chloride, potassium, and some two dozen other chemicals that make up the human body. In reporting these findings Professor T. Rivial commented, "It does make one humble."

The names of the professor and the university change annually, and the figure for the cost gradually rises with inflation, but otherwise the story remains the same. It's a damned silly story. The silliness would be immediately apparent if we converted it into the following:

> According to scientists at Jeremy Bentham University, the value of Rembrandt's painting "The Anatomy Lesson of Dr. Nicolaes Tulp" is $4.27. The estimation was made by a spectroscopic determination of the amount of iron, copper, cobalt, uranium, and other elements in the famous painting.

Research that assigns a monetary value to a human being or an oil painting as determined by the chemical elements alone is plainly an example of numeracy gone berserk. The desire to know the value of things is understandable, but answers like these are ludicrous. What's wrong?

The Conservation of Physical Wealth

Only three categories are required to classify all the many forms of worldly wealth: *matter*, *energy*, and *information*. The order given is also the historical order in which these categories became clear to science. The most critical question to be settled for each category is this: *Is it conserved*? That is, does the amount remain the same through all permutations of state, neither increasing nor decreasing?

The answer is often not obvious. If you were to burn Rembrandt's painting, even if you cared little about art, you would feel that something was being destroyed. The canvas and the paint would disappear. However, were you to carefully capture and weigh all the products of the combustion (principally carbon dioxide and water), you would find that they weighed more than the painted canvas that had disappeared. (This is because you neglected to weigh one of the raw materials, oxygen, which entered the system unnoticed from the air.) Lavoisier was the one who taught chemists the meticulous methods of accounting for matter that are needed to say whether matter is, or is not, conserved during chemical reactions.

By the middle of the nineteenth century the implications of Lavoisier's work had been converted to dogma: *Matter can be neither created nor destroyed.* This is the Law of the Conservation of Matter. A short while later a comparable Law of the Conservation of Energy was put into words: "Energy can be neither created nor destroyed." It is also called the First Law of Thermodynamics.

Energy was a bit harder nut to crack than matter. It is necessary to distinguish between what we may call "total energy" and "useful energy." Total energy is totally conserved; useful energy is not—some of it is lost in every change of state. For instance, water flowing from a dammed lake through a turbine can be used to generate electricity, which can then be used to run a pump that lifts the water coming out of the turbine back up to the lake: but the amount of water lifted up will be less than the amount that came through the outflow pipe lower down. In each such cycle, part of the "head" of water will be lost until the system finally shuts down, leaving all the water at the lower level.

Only if you carefully measure the heating up of the moving parts (bearings, et cetera) and the almost imperceptible heating of the large body of water, can you verify that energy is indeed totally conserved in this system. But it takes more than conserved energy to run a machine: there must be a difference in energy levels, as there is between the water in the higher "source" (dammed lake) and the lower "sink" (turbine outflow).

It is the difference between source and sink that runs our machines. Such differences are being continually degraded. *Useful energy decreases over time*—such is the Second Law of Thermodynamics. This is a law of *non*conservation.

This law can be stated another way. The physicist's word

"entropy" stands for the degree of disorder in a system. The difference between source and sink constitutes order, of a sort. "Entropy always tends to increase toward a maximum"—this is another way of expressing the Second Law. (In our example, entropy would be at a maximum when all the water was at the level of the turbine outflow and incapable of doing any more work.)

Textbooks on thermodynamics deal with all this in a much fancier way, with partial differential equations and the like. For our purposes a humorous version of the laws of thermodynamics suffices to show the human significance of these basic laws:

> You can't win;
> You are sure to lose; and—
> You can't get out of the game.

What double-entry bookkeeping is to business, the conservation laws are to science: they are essential to honest knowledge. New products or new methods may permit a business enterpriser to make more money than before, and new machines may operate more efficiently than the old, but never are the conservation laws violated.

With the coming of the twentieth century, some new findings in physics appeared to undermine the conservation laws. Radium and a few other substances were found to produce heat spontaneously (or at least apparently so). This was a serious threat to the sanity of physics. Then Einstein, in the year 1905, brought together the two previously separate conservation laws to form his single law, $E = mc^2$. Einstein found the factor that relates the amount of energy to the mass of material that produces energy when it disintegrates. The community of physicists breathed freely once more: conservation had been restored.

(Note in passing: Although Einstein united matter and energy into a single equation, most everyday affairs are conveniently treated as if the conservation of matter and the conservation of energy were separate matters. They will be so treated here.)

Wrestling with Information

In the public arena the conflict between conservationist and non-conservationist views continues to this day. It is a conflict in which there is seldom any joining of the issues: it is as if one side were speaking of apples, the other of oranges. As far as matter and energy are concerned physicists are confident that conservation rules everywhere. What is not conserved is something that is neither matter nor energy, which is variously called "organization" or "information." It was not until the middle of the present century that scientists gained a clear understanding of this category.

Information may be "coded" in either matter (print on a page) or energy (radio pulses). A minimum amount of one or the other is required: you can't encode the encyclopedia with a single atom of matter or a single quantum of energy. But beyond the required minimum there is no quantitative relationship between the amount of coded matter or energy and the "quantity of information" conveyed. The Lord's Prayer engraved on the head of a pin has no less, and no more, information than the same prayer writ large in the family Bible.

Multiple identical encodings do not increase the amount of information (though they do increase its availability, and hence its potential effect). If half the printed copies of *Hamlet* were destroyed, the "information" there encoded would still be with us. Only if literally the last vestige of *Hamlet* were destroyed—every last copy, computer disk, and mental record of the play—would there be a sudden quantitative shift from *Hamlet*-information to *no-Hamlet*-information. When dealing with information, the difference between 1 and 1,000,000 replicates is trivial as compared with the difference between 0 and 1.

Looked at from the other end, when someone writes a new play, discovers a new scientific principle, or invents a useful new machine, something seems to have been created. The very idea of creation contradicts the idea of conservation. Creating novelties is an ability of many kinds of animals, but *Homo sapiens* is supreme in this ability.

Condorcet, in hiding from the winning forces of the French Revolution, thought furiously about the future of the human species and came to the conclusion that "nature has assigned no limit to the perfecting of the human faculties, that the perfectibility of man is truly indefinite." This key passage, plus the title of his book, *An Historical Picture of the Progress of the Human Mind*, indicate that Cordorcet was thinking primarily of information rather than matter or energy.

It took another century and a half after Condorcet for the technical meaning of "information" to be clarified. Pundits of the nineteenth century quite unconsciously slipped from assertions about the accumulation of information in the human mind ("*l'esprit humain*," in Condorcet's words) to assertions about matter and energy. The nonconservation of information was unconsciously carried over into the material realm. The idea of progress became a very materialistic one, leading to the faith that technology would forever increase the amounts of energy and matter available to mankind.

Will this kind of material progress continue indefinitely? Or will the rate slow down until it is substantially zero? We have no firmly grounded theory to guide us to the answer. The number of scientists and technologists is still growing, and the bulk of scientific papers doubles about every ten years. But how much of the paper increase is a true increase in information? How much is attributable to scientists trying to make themselves heard over the roar of intellectual and organizational traffic jams? (After all, more gasoline is used in a big city than in a village, but much of it is a tax levied by congestion.) When traffic really jams, progress drops to zero. A substantial *and growing* proportion of the scientific literature is pure jam, the consequence of egotistic scientists putting out multiple, repetitive publications in an effort to be noticed. Jam grows by positive feedback as the number of scientists increases. Progress is impeded. Society suffers.

Computers are a great help in dealing with mountains of data, but there are limits even to this help. There are at this moment unbelievable quantities of data in computer banks—e.g., information sent back from satellites—that will probably never be "read" in any meaningful sense. (Scientists neglect to mention this when applying for new grants.) Information is a key factor in the progress of science. In the end, constipation may be as lethal as censorship.

Nuclear Energy: Escape from Conservation?

Because energy is conserved, and useful energy is steadily lost, no perpetual motion machine is possible. Scientists stopped looking for one in the nineteenth century, and the U.S. Patent Office stopped accepting applications for such machines in the twentieth. A few bemused amateurs still keep plugging away.

During the 1950s the vast amount of energy extractable from only a tiny amount of matter in a nuclear reactor created a euphoria in which the spirit of conservation was lost. One of the most influential spokesmen for the new euphoria was Lewis Strauss, an investment banker who was appointed chairman of the Atomic Energy Commission. Promoting nuclear energy before the National Association of Science Writers in 1954, Strauss said:

> It is not too much to expect that our children will enjoy in their homes electrical energy too cheap to meter, will know of great periodic famines in the world only as matters of history, will travel effortlessly over the seas and under them and through the air with a minimum of danger and at great speeds, and will experience a life span far longer than ours. . . . This is the forecast for an age of peace.

An investment banker is no doubt very good at numeracy in the field of investments, but his numerate ability does not necessarily carry over into science, technology and the ecolate interactions of human beings. The very great costs of the research that made atomic power practical were commonized over the national treasury; much of the cost of constructing nuclear-energy plants was also commonized; and the Price-Anderson act freed the energy establishment of all liability for nuclear accidents beyond the $560 million level, which meant that the cost of really big accidents would be commonized over the whole nation. No wonder the prestigious chairman of the A.E.C. convinced so many people that electrical energy would soon become "too cheap to meter." He even put his faith in the atomic future on a religious basis: "My faith tells me that the Creator did not intend man to evolve through the ages to this stage of civilization only now to devise something that would destroy life on this earth."

Designated as promoter of an exciting new technology the economically trained investment banker apparently forgot the greatest conservation principle of economics, "There's no such thing as a free lunch." He thought that his children would have almost free power; instead, in 1979 they got Three-Mile Island. Official calculations have never made adequate allowance for the changes that time brings to the strength of materials under ceaseless atomic bombardment or for the inescapability of human-generated failures under the bombardment of boredom or sabotage-generating disaffection. The total cost of nuclear power is unknown, perhaps unknowable.

Implications of Nonconservation

Yet still there are optimists who say, "Perhaps science will invent a free lunch tomorrow! After all, who could have foreseen the ultimate utility of electricity the day Franklin flew his kite?"

Who indeed? There is no definitive reply. The future is unpredictable. Santa Claus may exist after all. The crucial question is this: Does the prudent person base his plans on the faith that Santa Claus exists?

Scientists have no doubt that the prudent course is to assume that the laws of conservation of matter and energy will continue to hold. But *information is not conserved*, and that fact has a number of interesting consequences that are not widely enough appreciated.

Because matter and energy are conserved any system of distribution of these forms of wealth is a Zero-Sum Game. What I gain, you lose; and vice versa. Redistribution does not create wealth.

(Not directly, at any rate. Human affairs involve more than the simple laws of physics; for instance, they involve complex questions of human motivation. The redistribution of the physical wealth of farmlands in a poor and politically backward country may stimulate farmers to produce more abundant crops. Such effects clearly lie outside the realm of physics. Awareness of such trans-scientific effects makes it hard for some people to think straight about scientific conservation laws.)

The distribution of information is not a Zero-Sum Game. More often than not it is a Plus-Sum Game. If I have a bright idea, and if I share it freely with others, some bright person may work over my idea and pass it back to me in improved form. That's a Plus-Sum Game; the history of science is replete with examples. This is why science is best pursued in a completely open manner. Science is international; or supranational, if you wish. This is its glory and its strength. To restrict science is to weaken it.

The same principle applies, though perhaps with less force, to all information, scientific or not. It is the literate world that carries out most of the redistribution of information, so it is not surprising that professional literates are, as a class, great supporters of "freedom of information." At the fringe of their empire, e.g., when it comes to pornography, the unqualified condemnation of restriction (censorship) leads to policies that not all people are pleased with. But over most of the realm of literacy, freedom of information is accepted as the best policy.

The herdsman's concept of "carrying capacity" has no meaning applied to the wealth of literature, which can be shared widely without loss—indeed, with positive gain. This fact of his profession may make it easy for a wordsmith to be taken in by proposals to globalize the wealth of materials and energy, commonly through an initial step of globalizing a problem. Globalizing material wealth means creating a material commons, yet many a literate professional sees nothing wrong with the axiom of the nineteenth-century anarchist, Pierre-Joseph Proudhon, who said that "Property is theft." Under this rubric legal thievery becomes no more than a timely redistribution of wealth logically justified by a human rights doctrine.

Dilemmas of the Literate

Professional writers are caught on the horns of a dilemma. On the one hand they want their creations disseminated as widely as possible; on the other hand they are willing to restrict distribution when that is the most effective way to bargain for payment for the effort that went into their work. Commonization of his ideas is the dearest wish of the wordsmith; he is less enthusiastic about commonization of the material rewards of his creativity.

Copyright law makes it possible to privatize something that is, by its nature, highly susceptible to commonization. In the nineteenth century Charles Dickens and other prestigious literary figures successfully fought the battle for the privatization of literary property. Because of them it is now easy and almost automatic to copyright a book in most of the countries of the world. As of the first half of the twentieth century the battle seemed to have been won.

Then in the 1950s the Xerox machine was perfected. In the early days of xerography the new technology did not seem much of a threat to the copyright system, for a single xerographed sheet generally cost ten cents. Professional books were selling at this time for about three cents a page. In subsequent years the price of xerography went down while the price of printing went up. At the present time, it is not unusual for textbooks and professional books to cost fifteen cents a page. Even higher prices are known. In contrast, xerographed sheets now cost four to six cents apiece, and often two pages of a book can be copied onto a single sheet.

Though a bound book is easier to handle than a bunch of loose sheets, a price differential of a factor of four or more creates a great temptation to make unauthorized copies. The temptation is seldom resisted. As for journal articles, a stapled copy of a single article may actually be more convenient to read and store than the whole issue of a journal, which may include twenty unwanted articles. Some important scientific papers now exist in greater numbers in xerographed form than in the original print.

Xerography has effectively created a new commons. Literate professionals of Marxist bent who promote distribution according to need should wholeheartedly praise this new development. But, as Martin Luther said more than four centuries ago, "It makes a difference whose ox is gored."

Many attempts have been made to control xerography. Copy shops display prominent signs warning against the reproduction of copyrighted material, but how often does the proprietor of the shop look to see what his clients are copying? Trying to limit the commons of xerography, the editors of many scientific journals set a price for reproductions at the bottom of the first page of each article. The client blithely xerographs the price along with everything else.

Expensive musical recordings are copied endlessly on less expensive tapes. Disks bearing software for computers are flagrantly copied. Here the cost differential between copy and original may be enormous: five dollars versus five hundred dollars or more. That's a lot of temptation.

The idea behind the copyright law is not just to protect the interests of the creator so that he can get rich. It is also in the interest of society that potentially commonizable material be privatized, at least for awhile, so that other ingenious people will be motivated to produce in the future. Great rewards are great motivators. Technological advances in the replication of information at low cost have created a new form of commons that is hard to fence in. Technology seems to have thrown morality for a loss.

Pure Research: A Problem for Society

In science, pure research and applied research form a continuum. At the applied end it is easy to generate economic motivation: patents and royalties hold out the promise of fortunes. But pure research is, by definition, investigation at so fundamental a level that it is unlikely to yield a financial payoff for many years, perhaps for generations. In the long run, applied research (called "Research & Development" or "R & D") depends on pure research, but the payoff for the pure can be painfully long in coming—too long for a small but well run business to risk financing pure research. Only a handful of our largest corporations (e.g., General Electric and American Telephone & Telegraph) have been able to support a significant amount of pure scientific research. (What will happen to the distinguished work carried out in Bell Labs now that AT&T has been broken up is an interesting question.)

The costs of pure research in this generation may be more than matched by benefits in generations to come. But the double question "Who benefits? Who pays?" demands a double answer. Only this generation is capable of paying, though the benefits go to people who are incapable of paying because they do not yet exist. When the beneficiaries finally come into being they will find it impossible to repay their benefactors, who will have long been dead.

A society of "bottom-liners," wholly oriented to the present, cannot justify the support of pure research. Then where can this support come from? Until this century many, perhaps most, pure scientists have been amateurs in the strictest sense—literally, "lovers" of learning, not people paid to devote their lives to study and investigation. Many of them (e.g., Darwin, Cavendish) were men of independent means. Others held jobs in unrelated fields that imposed such light demands on a person's time that he could turn out first-rate science in odd hours (e.g., Einstein, employed as an examiner in the Swiss patent office). In addition, research in earlier days did not require the expensive equipment it does now. First-rate research could be carried out with only marginal support by society.

Now society has to pay big for pure research. But who is "society"? Since few industrial concerns can afford such philanthropy, most of the support has to come from the nation as a whole. The time between investment and payoff is so long, and the connection between the two so tenuous, there is little chance of economically tying one to the other. And, as we learned with nuclear energy, the secrets of nature cannot be kept locked up by those who paid for their discovery. This means that pure research presents us with a new distribution system: Privatized Costs and Commonized Profits—with the commons extending over the whole world (or at least that part of the world that is well enough educated to understand the results).

Those who dream of One World, a world of commonized wealth, should know that, with respect to information, such a world is already in being. Though friction in the distribution system (e.g., patents, copyrights, and censorship) can delay the commonization of new information for awhile, in the long run information is passed around among all those who are capable of absorbing it, whether their intentions toward the rest of the world are benign or malign.

Civilization can be viewed as a manifestation of accumulated information. Is there a law of conservation for civilization itself? Only an incurable believer in Providence can suppose there is. The destruction now being wrought by a few saboteurs and terrorists precludes such simple faith. Entropy, which needs no helpers, has all too many in our time. With the continued growth of population in our finite world the number of entropy's helpers will increase more.

The nonconservation of information works both ways. Increase is as possible as decrease; both destruction and creation are possible. As Condorcet dimly realized, there are no assignable limits to improvements in information and organization. Even if there are practical limits to the material wealth available to human populations, which have a potential for limitless growth, the most important elements of that difficult-to-define entity we call "the quality of life" are not material and hence may not be limited. After nearly two centuries of mistakenly pursuing limitless material progress, it is time that we shift the major focus of our attention to the immaterial aspects of life. Above all else, this means we must bring human demands into balance with material resources. Though these are limited, human ingenuity in tackling organizational problems knows no certain limit.

CHAPTER 17

One World: An Ecolate View

Long before written history began, men must have known that local action was far more likely to succeed than action directed from a distance. Yet localism—the parochial orientation—was constantly eroded by the necessities of warfare, which (in general) favors larger groups. The larger size brought into being by war was, quite naturally and despite occasional criticism, carried over into peace. The resulting overgrowth of the civilian state entailed a loss in operational responsibility, but this disadvantage was counterbalanced by an increase in the survivability of the group under conditions of war (which was never far off). Power pays.

For most of recorded history most people's commitment of affection and respect for groups stopped far short of all humanity. Yet for the past two thousand years isolated seers and saints have deplored the narrower sympathies. It was not until the twentieth century that there was a mushroom-like growth of the universalist sentiment, resulting in such statements as the following:

- "To see the earth [from space] as it truly is, small and blue and beautiful in that eternal silence where it floats, is to see ourselves as riders on the earth together, brothers on that bright loneliness in the eternal cold—brothers who know now they are truly brothers."
- "On ecological grounds, the case for world government is beyond argument."
- "The historic doctrine of national sovereignty has been eroded

by the seminal truth that all contemporary problems, from mass unemployment to terms of trade and commerce, are global in character and in the long run can only be resolved by global forms of cooperative planning. The world is the only viable unit of reference."

- "Modern technology has rendered national sovereignty obsolete."
- "Apparently you cannot convince nations that they should assist others voluntarily. But people should realize that if no solution is found, the future looks rather bleak. If the rich countries will not share their wealth, the poor people of the world will come and take it for themselves."

These assertions were made by a poet, a political scientist, a sociologist, a philosopher and an economist. I do not give their personal identifications because the statements are unexceptional; similar sentiments have been voiced by scores of commentators on humanity's situation. Further discussion will be focused on two questions: Are the sentiments new? Are the implications, both logical and rhetorical, true?

Biblical Evasion

The ethical underpinning of the antinationalist quotations is of course the idea of being a "Good Samaritan," which comes to us from the Bible. Millions of Americans have, during their Sunday school days, been exposed to the account of the Good Samaritan, but if they have never read the Bible slowly and thoughtfully as an adult they may be surprised at what it does and does not say. In the following quotation from the Goodspeed version of *Luke* 10:25-37, italics have been added to call attention to the crucial question.

> Then an expert in the Law got up to test him [Jesus] and said, "Master, what must I do to make sure of eternal life?"
> Jesus said to him, "What does the Law say? How does it read?"

He answered, " 'You must love the Lord your God with your whole heart, your whole soul, your whole strength, and your whole mind,' and 'your neighbor as you do yourself.' "

Jesus said to him, "You are right. Do that and you will live."

But he, wishing to justify his question, said, "And *who is my neighbor*?"

Jesus replied, "A man was on his way down from Jerusalem to Jericho, when he fell into the hands of robbers, and they stripped him and beat him and went off leaving him half dead. Now a priest happened to be going that way, and when he saw him, he went by on the other side of the road. And a Levite also came to the place, and when he saw him, he went by on the other side. But a Samaritan who was traveling that way came upon him, and when he saw him he pitied him, and he went up to him and dressed his wounds with oil and wine and bound them up. And he put him on his own mule and brought him to an inn and took care of him. The next day he took out two dollars and gave it to the innkeeper and said, 'Take care of him, and whatever more you spend I will pay you for on my way back.' Which of these three do you think proved himself a neighbor to the man who fell into the robbers' hands?"

The "expert in the Law," recognizing his cue, then says "The man who took pity on him"—and there the Biblical account ends. Yet if the expert had a keen mind he must sooner or later have realized that he had been bamboozled. The person who helps *me* is easily recognizable as a good neighbor—an easy answer to an easy question. The hard question is, whom should *I* help? Jesus sidestepped this question.

The question cannot be answered without considering either numbers or discrimination. Am I obliged to help everyone in the world, no matter how numerous? If the answer is "Everyone," then the discussion is at an end, but the task is impossible.

If, on the other hand, the answer is "Not everyone," then

we have to pass to the next question and ask, "On what basis do I discriminate between the ones I help and the ones I pass by?" Do I make my choices on the basis of distance in space, in religious beliefs, or in ethical commitments? What are the boundaries of the class called "neighbors," whom the individual should love as himself? Competent publicist that he was, Jesus evaded the question he did not want to deal with by offering an answer to a related but only partially relevant question. Good theater, but it does not advance knowledge. A great deal of formal ethics is clever evasion. All that many people want out of an ethical discussion is a warm feeling.

At this late date in the post-Christian Era, it might be questioned whether the Bible should even be brought into an ethical discussion. I argue that it should. Many people who do not explicitly call on the Bible for backing are nevertheless influenced by what they believe the Bible says. (Everyone knows about the Good Samaritan, even though they may not be able to remember what Samaritans in general were.) It is considered good form to speak politely of Scripture: revere it, but don't bother reading it seems to be the rule. Above all, don't read it thoughtfully.

Zeno to Marx

Christianity has important roots in the doctrines of the Stoics in the third century B.C. The Stoic philosopher Zeno of Citium (not to be confused with the mathematician Zeno of Elea) introduced a new view of commitment. Denying any overriding loyalty to his city (*polis*, which then meant city-state), he identified himself as a "cosmopolitan"—a citizen of the world (*cosmos*). His views unquestionably influenced St. Paul and later Tertullian; and through these men, all Christianity.

The cosmic outlook was greatly advanced by Marx's "From each according to his ability, to each according to his needs."

This directive makes no mention of tribal, religious, or national limitations to distribution systems. Indeed, communism has been an international movement from the beginning and has thus fitted into the panglobal orientation of the Christian missionary movement of the nineteenth century. Considering Marx's antipathy to all religions—"Religion is the opium of the people" are his words—it is a delicious irony that he succeeded so well in promoting a major goal of Christianity. The similarity in aims of Christianity and Marxism helps explain the phenomenon that the self-taught philosopher Eric Hoffer pointed out in his book, *The True Believer*—the ease with which a passionate Christian can become a passionate Marxist, and vice versa.

Singular Ethics

On the scale of a family or a small village the Christian-Marxist ethic may work well enough; but long before it can be extended to include the nearly 5,000,000,000 people of our earthly cosmos it fails because "according to his needs" operationally defines a commons, which necessarily leads to the tragedy of the commons. Inevitably. Why, then, do people continue to support this patently unworkable ideal? How could the mistake have been made in the first place?

We should note that Zeno dealt with words (city, citizen, cosmos) not with matter and energy. Matter and energy obey conservation laws, words do not. Jesus, St. Paul, Tertullian, and Marx were all wordsmiths and, in their professional work, could afford the cosmopolitan outlook. Of course the ready defense of wordsmiths is that they are "intellectuals," and as such they merely put into exact words the ideals that all must follow. But the work of wordsmiths is worth little if it is not first passed through the gauntlet of critics skilled in numerate and ecolate analysis.

Let's look at Scripture once more, this time passing it through the numerate filter.

"*Which of these three do you think proved himself a neighbor to the man who fell into the robbers' hands?*" Numerate note: Neighbor is singular, and man is singular.

"*Am I my brother's keeper?*" Numerate note: Brother is singular.

"*A new commandment I give unto you. That ye love one another . . .*" Numerate note: One is singular.

The italicized statements implicitly define what can only be called *singular ethics*, the belief that whatever is proved for the singular holds also for all larger numbers, without limit. Singular ethics disdains numeracy and is contrary to the spirit of science. Engrained in the very marrow of science is the abiding belief that

> numbers matter;
> quantities matter;
> ratios matter.

Mathematics possesses a powerful method known as "mathematical induction," whereby one proves the generality of a theorem by showing that whatever is true for n is true also for $n+1$, without limit. Mathematics can do this because it abstracts numbers from all else. In the material world, however, associated properties cannot be separated from one another. As the number associated with the size of an object increases, all the properties of the object change too, but the way one property changes may follow a mathematically different law from that of another property change. As the scale of an object changes the ratios of its properties change. The result is the scale effect, already discussed in Chapter 12.

Men cannot fly—not in the sense that hummingbirds fly. Bryan Allen, the first man ever to fly by human muscle power, weighed 62,597 grams. Working furiously in engineer Paul MacCready's 31,752-gram *Gossamer Condor*, Allen just barely managed to fly. There is no way in the world that an adult human being could ever fly with the breathtaking precision and flexibility of a hummingbird, which weighs only 3.5 grams. Most men and women, weighing much more than Allen's 138 pounds (and being less well muscled), have not the ghost of a chance of flying through the air. Scale matters.

Am I my brother's keeper? If he is only one brother, singular ethics may apply. But if "he" is two billion brothers and sisters—the approximate number of people in the world who are below America's official "poverty line"—singular ethics becomes a recipe for ruin.

It is not only scale that undermines the feasibility of conventional global ideals: rates do also. The rate of reproduction of the needy is much greater than that of those asked to be need-suppliers. The poor of the world typically increase at 2 to 3 percent per year. The comparatively rich—which includes almost everybody in the United States, even though most of them feel they are far from truly rich—typically increase by less than 0.5 percent. While the desperately poor are increasing by 40 to 60 million each year, the less than a billion people who are modestly well off are increasing by no more than 5 million. As a result, the proportion of the truly needy grows each year.

Little good can come from the practice of singular ethics, ignoring as it does numbers, scale, and rates.

Replacing Abstractions with Actors

The record of what poor countries have done with donations from rich countries does not encourage us to believe that giving is the way to end "global poverty." A particularly outrageous example of the way foreign gifts may be used by recipients has been given by Daniel Patrick Moynihan. The place was Gabon, a West African nation of less than 600,000 people. The occasion was a reception for the Organization of African Unity, given by President El Hadi Omar Bongo in July 1977.

President Bongo prepared a truly gala affair for his fellow leaders. He had built a reception hall in Libreville for some $250 million, including 52 villas for the delegation heads, two swimming pools, a luxurious night club, a sauna, a gymnasium, and two theatres. The complex had a unique feature: a viewing room between the two theatres which had a rotating floor, so that President Bongo could watch either stage by pushing a button. He did not have to go to the inconvenience of swiveling his chair. According to *The New York Times* account, the delegates to the conference were escorted from the airport by phalanxes of motorcycle outriders in frock coats (the delegates themselves rode in armored Cadillacs), and the parade route passed by guards in plumed kepis and crimson robes, as well as troups of singing and dancing women wearing T-shirts with President Bongo's picture on them.

This, in a country with a national debt of one billion dollars, the interest payments on which were eating up 23 percent of the annual budget. Conspicuous consumption on this scale is theatrical, but the aggregate waste from less spectacular activities is probably greater.

A proposed gift from rich country to poor country may run into opposition in the rich country, in which case the transfer of funds may be written up as a loan. With the passage of time the borrower may fall behind in the payments, at which point the loan is likely to be converted into a gift. The realistic view is to regard the transfer as a loan from the beginning. When the deal turns sour, we should blame the donors more than the recipients. The donors should have detected the reality behind the facade of international charity. No doubt the donors' thinking runs something like this:

a. The people in country X are poor.
b. If we give enough to country X there will no longer be poor people in country X.
c. Therefore let us give generously to country X.

But there is a great gap between "country X" and "the poor of X." How is this gap bridged? Money given to the government of X is distributed by the people who run the government, and these are not the poor people whose plight tugs at the heartstrings of rich foreigners. The governors of X distribute most of the donated funds in the following ways, which are not mutually exclusive:

a. For prestige items, e.g., a national airline. I believe every African country south of the equator has its own airline, but precious few of the pilots are natives.
b. To enrich the families or tribes of the rulers, in keeping with unbroken tradition which is strongly tribalistic and family-oriented.
c. And for graft of many sorts, which shocks the donors more than the recipients. In Spanish-speaking countries graft is called *la mordida*, "the bite," and is regarded much as we regard a commission.

Country X, which is an abstraction, cannot act. Those who act in its name are rich and powerful people. Human nature being what it is we can be sure that these people will not voluntarily do anything to diminish either their power or their riches. Prudence may move the rich to let a few goodies trickle down to the poor, but the trickling will not be enough to undermine the favored position of the rich. How could it be otherwise?

Gross as it is, the wastage due to what external philanthropists view as injustice is often less than the waste due simply to trying to do too much too fast. Poor countries are notably lacking in "infrastructure"—good roads, railroads, trucks, adequate storage facilities, and the trained personnel needed to operate the infrastructure. Not uncommonly, the major part of large quantities of food sent in haste to a poor country in the tropics rots on the docks or is eaten up by rats before it can be moved to the people who need it. The wastage is seldom adequately reported back to the sending country, because few people stand to gain personally from publicizing an unpleasant reality. Those who gain personally from the shipping of food to poor nations gain whether fungi, rats, or people eat the food.

Philanthropy that truly benefits the recipient is not impossible, but it requires close attention and postaudits of the results, so that donors can learn from their mistakes. It is easier to make material gifts than it is to garner information and learn from it.

CHAPTER 18

Telephilanthropy: A Modern Invention

Who first said, "Charity begins at home"? Bartlett's *Familiar Quotations* credits Sir Thomas Browne, whose *Religio Medici* was published in 1642. The thought surely was not original with Sir Thomas. Selection theory, as developed by biologists, convincingly shows that discrimination must have been a part of charity from the very beginning, long before our proto-human forebears either spoke or wrote. Practicing charity at a distance was so bizarre an idea to our ancestors that it did not even need refuting until the seventeenth century. Be that as it may, it is certain that telephilanthropy (to give the practice a name) is essentially a modern invention.

Giving makes the giver feel good: for the mine run of "do-gooders" this is justification enough for the act. Contentment is harder to achieve among the minority of philanthropists who are not satisfied unless a postaudit shows that their acts have in fact benefited the receiver. Men who have both made and given away millions testify that giving intelligently is much more difficult than making a fortune. We who are not rich may find that hard to believe, but we should be impressed by overwhelming agreement among those in a position to know.

It is distressingly difficult for an individual donor to do good. For a nation, doing good is even more difficult. The multiplicity of opinions and interest groups among both donors and recipients complicates the making and carrying out of plans. In earlier days when most of the donor groups were individual churches, agreement on goals was easier to reach than it is now when the donor is a pluralistic nation.

The nineteenth century saw a great burgeoning of missionary philanthropy. To modern, psychologically sophisticated people the primary motive of this effort may be in doubt. Was it to earn a place in heaven for the parishioners? Was it to help the heathen? Was it to give employment to agents of the church? Or was it to help the sect to outgrow its competitors? In any event, material aid was closely tied to religious conversion; this was in keeping with the spirit of St. Paul's Epistle to the Galatians.

St. Paul did not speak of a "brotherhood of man," but only of a brotherhood of believers in Christ. Leaders of other religions have been equally discriminating. It is only in the past century that the ideal of a religiously unconditional brotherhood has gained much of a following.

(Note in passing: To be thoroughly fashionable we should now speak of a "siblinghood of persons." Schiller's affirmation, "*Alle Menschen werden Brueder*," also needs revision.)

The Marshall Plan

Telephilanthropy on a substantial scale was impossible until the development of modern means of transportation and communication. Even with these improvements it was not obvious that improving the well-being of foreigners was in the national interest. Until the twentieth century telephilanthropy was practiced only by what we now call NGOs (Non-Governmental Organizations).

The end of World War II brought a radical change in outlook. This took place in two steps. The first step was the Marshall Plan. As soon as the war ended, it was obvious that the wartime cooperation of Russia and the United States was changing into an intense competition for the political loyalty of European countries. It was widely held that poverty, disorder, and suffering favored communism over democracy. Paradoxically, Americans, so optimistic about the outcome of competition between Western technology and other ways of dealing with the material world, easily subsided into pessimism about the consequences of competition at the ideational level between communism and capitalism. America's response to the perceived threat was to try to rebuild Europe as fast as possible. Secretary of State George C. Marshall argued forcefully for this course of action in a public address given at Harvard in June 1947, and his plan was soon adopted.

The Marshall Plan is a rare instance of a large public operation that succeeded better than its most enthusiastic supporters expected. Over a period of four years the United States poured some seventeen billion dollars into the Plan. European governments spent somewhat less, but their people contributed enthusiasm and the local know-how required to convert money into action. The physical structure and the organizational infrastructure of European civilization was recreated and even improved. Ironically, some of the "conquered" countries of Europe seem to have emerged from war-plus-reconstruction in better shape than the nominal victors.

The philanthropic element of the Marshall Plan was incidental to its main purpose, which was to create a Europe strong enough to act as a bulwark against the Communist threat from the east. Because the principal goal was defense it was easy to gain the assent of the American electorate to the drain on the common wealth of the nation. We were, as was said from 1947 on, engaged in a "cold war." The justification for dipping into the national commons was the same as it is for all wars.

Because Marshall funds were earmarked for the joint defense of the Western nations it was possible to keep them from being dispersed "to each according to his needs." The welfare needs of individual Europeans were acute, but at the insistence of the American overseers Marshall funds were restricted to the rebuilding of European industry.

Though the required study has never been made—and probably never will be—there is little doubt that many Europeans died because Marshall money was channeled into industrial hardware instead of being spent for food, clothing, and fuel for heating homes. The justification of this choice is simple: many more people would probably have died within a few years had the immediate creature needs of Europeans been met first, thus neglecting the rebuilding of the industries required to take care of whole nations.

As a result of centuries of experience it may be that Europeans are more willing than Americans to grant the priority of group interest over individual interests or "rights." In the postwar reconstruction they acceded to sacrifices that would shock many Americans. During one of the coldest winters on record, Vienna refused to let the famed Vienna Woods be harvested for fuel. And as soon as they could, the Viennese allocated funds to rebuild their lavish opera house, which had been destroyed in the war. In the Viennese economy, music is more than a luxury: it is an industry.

The "Bold New Plan" of Foreign Aid

Great enterprises, like human babies, are sometimes conceived almost absent-mindedly. When Harry Truman achieved his surprising victory in the presidential election of 1948, he started looking around for a "gimmick" to make his mark on history. His first problem was what to put into his inaugural address.

Truman's attention was soon called to a suggestion that had been bouncing around the State Department for some time without getting significant support. The idea was to use American dollars and advice to facilitate "development" in poor countries that seemed to be getting nowhere in their struggle against poverty. The idea had no name, but Truman liked it and it ended up as the last of four points in his address, hence its first (and least informative) name, "Point Four." It was not until June 1950 that Congress adopted the program and gave it some funds ($45 million initially), calling it the Technical Cooperation Administration. Another three years passed before the program was christened "Foreign Aid."

Truman was right when he called it a "bold new plan." It was, as he said in his memoirs, "an adventurous idea such as had never before been proposed by any country in the history of the world." Since there is no assurance that intervening in the lives of others will actually aid them, the most objective name for Truman's program would have been "Foreign Intervention," or perhaps "Non-Bellicose Foreign Intervention." But language is action, and names are chosen to keep auditors from questioning assumptions. The prejudicial term "Foreign Aid" does just that.

Has foreign intervention brought genuine aid to foreign countries? The answer depends on whom you ask. The bureaucrats who administer Foreign Aid of course assert its effectiveness. But you should never ask the village barber if you need a haircut.

Observers with no axe to grind give a mixed reading. Here and there intervention has truly helped, but quite often what Westerners call "development" has proven to be damaging to societies of a different sort. During the 1950s economists convinced the foreign-aid establishment that the motto should be *Industrialize!* When a decade's experience showed that this procedure did not work in most places a new motto was adopted: *Mechanize!* This meant that "primitive" agriculture should be mechanized. (In the 1960s economists recommended that India's water buffaloes be replaced by tractors. Then came the OPEC-generated oil crisis of 1973. . . . Back to the drawing boards once more!)

An answer that had been tried out for a number of years by the English economist E. F. Schumacher was published in the year of the oil crisis in a small but immensely influential book, *Small Is Beautiful.* The subtitle touched people's hearts: "Economics as if People Mattered." Schumacher's message was *Appropriatize!*, that is, introduce measures that are appropriate to the existing life circumstances of the recipients.

The foreign aid effort is in many respects a thoughtless rerun of the nineteenth-century missionary story. I recall one conference at which a staff member of the Senate Foreign

Relations Committee, speaking of the poor of Jamaica, said: "They have a problem and we have to begin tackling this problem." I asked her to repeat, and she did. My ear italicized the *they* and the *we;* I was struck speechless by the linguistic legerdemain. The speaker was apparently unaware of the paternalism implied by the shift in pronouns, and without doubts as to the wisdom of the shift in responsibility.

Psychologically, foreign aid in the twentieth century draws on the religious enthusiasm that propelled the nineteenth-century missionary enterprise, but the gods are now secular gods: "Progress" and "Development" are their names, where it is understood that we mean technological progress and technological development. Implied, and sometimes said, by enthusiasts for technological progress is the doctrine, "Bigger is better." Schumacher's "Small is beautiful" was a welcome antidote to that kind of thinking. The progress needed in a poor country is more subtle than the technological kind the Western world worships. Granting that, we must then face the next question: Just what kind of development *is* appropriate? There are no easy answers.

But if we are uncertain, are we justified in intervening? What if intervention does more harm than good? Those who are interested in following up this question would do well to start with the writings of the English economist P. T. Bauer.

A complete evaluation of the multitudinous foreign interventions is too ambitious for the present work. We would do well, however, to try to understand how America was persuaded to undertake a program that, to date, has spent approximately a hundred billion dollars. (This is the figure that is left after subtracting what is estimated to be the amount spent on foreign military aid, which is frequently camouflaged as Foreign Aid, a name that should be reserved to development aid and "direct aid," such as food and medical supplies. Because the camouflage is skillfully crafted, no estimate of the amount of true Foreign Aid is overly accurate.)

When first proposed, the foreign-aid bill looked like a tough one to get through Congress. In journalistic terms it looked like a case of taxing Americans to feed foreigners. That would be a simple Zero-Sum Game. But why should Americans play such a game? Out of generosity? Perhaps in part. But if that had been the only motive Foreign Aid would have died aborning.

The enabling legislation, principally Public Law 480, created a powerful constituency in favor of Foreign Aid. This was done by setting in place a Double C–Double P Game among our own people. Privatized profits coupled to commonized costs make a powerful steamroller.

Consider the consequences of sending food abroad under Public Law 480. American farmers did not *give* the grain to the world's needy: they *sold* it to the U. S. government agency that arranged for the gift—for which all the taxpayers paid. In other words, the cost was commonized.

Privatized profits accrued to those who owned and operated farms, grain elevators, trucks, railroads, and ocean freighters; and to a lesser degree to the bureaucrats who kept the system working. Those who made large profits could afford the costs of lobbying for Congressional support. Those who paid the costs—the taxpayers—were more numerous but were not organized and did not hire lobbyists. It was cheaper for each individual taxpayer to pay the additional taxes than to fight. The CC–PP Game worked its usual magic.

For those Americans who paid close attention to what was going on, Foreign Aid was a disillusionment. At the outset the program was supported by an argument that went like this: Four years and $17,000,000,000 spent on the Marshall Plan saved Europe—why can't a somewhat larger sum save the rest of the world? Alas, the thought was one more illustration of the wise saw that "All bad ideas are born good."

Whatever the motivation of the people who owned the farms, railroads and shipping lines, the idealistic language of the publicists was intoxicating. James Burnham, in his *Suicide of the West*, tells of hearing one of President Kennedy's young advisers say—his voice vibrating with emotion and his eyes shining—"So long as a single being anywhere in the universe suffers from hunger or any economic privation or any injustice, this nation has the duty to help him or her." Generosity? Immaturity? Lack of imagination? Hubris? Take your pick.

In any case, the experience with the Marshall plan furnished a misleading analogy for worldwide foreign aid. The analogy came naturally to people trained only in the use of the literate filter; it was not the sort of error that would be made by anyone who habitually passed reality through numerate and ecolate filters. Such critics would want to know the answers to a host of related questions. How many people are involved in the two kinds of programs? What percent of each group were literate? How well were they prepared by their history and culture for the making of great efforts? How much change was called for in their habitual way of looking at things?

Postwar Europeans faced a problem of *re*building a world they had once known; the poor elsewhere have to learn to do something that is utterly new to them. Europeans are of a Faustian temperament, holding to the engineer's faith that "The difficult we do immediately, the impossible takes a little longer." By contrast, most poor people are fatalists (for reason). Europeans are 99+ percent literate, with all that implies about education. Poor populations are typically only 5 to 30 percent literate, and that only by forgiving standards. ("Write your name. Good! You're literate.") Individualistic Europeans are self-starters; many poor people are pathologically passive.

Finally, the technological civilization of Europe had been brought into being during centuries in which the average growth rate of the population was well below one percent per year. Emigration to the New World was a marvelous safety valve for people who had not yet learned to control reproduction. That safety valve is almost frozen shut now. Populations in today's overcrowded nations are increasing at two to three percent per year.

To put the matter differently, Europe had the advantage of developing during an era when doubling the population took more than a hundred years; poor countries are now typically doubling every twenty to thirty years. Trying to develop into a modern industrial state under these conditions is like trying to work out the choreography for a new ballet in a crowded subway car.

Great Private Banks Embrace the Commons

If I owe the bank ten thousand dollars the bank controls me; if I owe a hundred million I control the bank. This is but one more example of the importance of numbers, magnitudes and relative size in determining what happens in the world. This numerate insight, generally credited to John Maynard Keynes, suddenly resurfaced in the fall of 1982 as the threat of default among borrowing nations led investigators to dig out the facts about the loans of private American banks to precarious national economies.

Brazil and Mexico had each been lent $68,000,000,000 by a blue-ribbon list of American banks (Chase Manhattan, Citicorp, Manufacturers Hanover and the like). With inflation running over 100 percent per year in Brazil, and the price of Mexico's oil plummeting, defaults were a real possibility. Since a single great default would threaten to topple the international monetary system like a house of cards, the borrowing countries were following Lord Keynes' implied advice and using the threat of default to secure more loans—to pay the interest on past loans! Like all forms of blackmail, this threat sets up a game with a tragic end.

The new development was a shock to a generation raised on the stereotype of the steely-eyed banker, realistic to the last about human hopes and plans. Of course we had come to expect more sentimental standards of the quasi-governmental World Bank and subsidiaries which had been set up specifically to make "soft" loans. As Charles T. Munger observed, "The World Bank is not a bank but a religion." But private banks had been presumed to be just banks. What had happened?

The watershed appears to have been the oil crisis of 1973. The dramatic rise in the price of oil channeled billions of dollars into the coffers of OPEC countries faster than they could spend them for outsize airports, high rise buildings and four-lane roads that led nowhere. Moreover the rulers of the sand-rich, oil-rich Arab countries realized that when the oil finally gives out about all they will have left will be the sand. Oil revenues have given them a once and fleeting chance to buy into enterprises in physically more favored parts of the world before the essential resource poverty of their native lands retightens its grip. So the oil magnates bundled up their billions and sent them off to Europe and the United States for investment.

A simple, direct response to the inflowing billions might have been to lower the rate of interest on loans. This would be in keeping with the philosophy of letting supply and demand create their own balance. But this, like many idealistic philosophies, is principally a product for export, something the speaker recommends for others but does not make use of himself if he can help it.

The thought of what a rapidly falling interest rate would do to the complex network of commercial relations was too frightening for American bankers to contemplate. Hewing to business as usual, the great commercial banks elected to lower their standards for making bank loans, while at the same time they maintained high rates of interest. They sought out borrowers in the poorest countries and made them offers they couldn't refuse. Since foreign aid was now so fashionable the new role gave private bankers a warm feeling that they were now part of the great religion called "interdependency." The full Keynesian meaning of the term only gradually dawned on them.

Were the private bankers naive? Or shrewd as foxes? Bankers are not given to writing their memoirs, so the general public will probably never know, but a plausible scenario would cast them in the role of foxes.

Once again we are confronted with a situation in which numbers matter. An increase in the magnitude of the borrowed sum relative to the resources of the lender finally causes a sort of dialectical reversal in power as the day of default draws ever nearer. At this point an international banker no doubt hopes that there will be available to him the international equivalent of our domestic Federal Deposit Insurance Corporation (F.D.I.C.).

When a regional bank fails in the United States the F.D.I.C. steps in to take over the assets and pay off the depositors. The losses incurred are then charged against the commons of the national treasury. This poses the usual dangers of the commons which, however, are lessened by considerable federal oversight of the lending behavior of private banks. So long as bank failures are relatively few the system works well enough. (What would happen if a majority of the private banks failed? Well, that is another story the ending of which we hope we never learn.)

There is no international equivalent of the F.D.I.C., no common pot to take care of massive defaulted international loans. It is, however, a plausible guess that the great international bankers, relying on the modern willingness to commonize disaster losses, are counting on persuading the government to create an ad hoc international F.D.I.C. that will take care of the American banks that carried out the essentially religious function of making shaky loans to poor nations.

Whether or not this is the correct explanation, the fact is that by late 1982 at least 73 percent of the loans to poor countries had been made by private banks, not by government or quasi-government agencies. We must say "at least," because records of private loans are not easy to uncover. Government loans are a matter of public record; private loans are not. On 18 December 1982, *The Wall Street Journal* estimated that at least $626,000,000,000 had been lent to poor countries by private banks—ample cause for anxiety in a world increasingly ruled by scarcities. Of course, if perpetual growth is our destiny we have nothing to worry about. *If* . . .

The Best Wordsmiths Money Can Buy

As the public took fright at the magnitude of the dubious loans made by the banks, bankers realized they had a job to do. They had to recreate public confidence in the international banking system. In banking, up until the very second when a government agent padlocks the doors, the officers of a failing bank have to exude confidence. Exuding confidence is something bankers are good at. Nowadays they are helped in this by "public relations" professionals.

No doubt helped by a PR man, a vice-chairman of the Chase Manhattan Bank gave a lengthy analysis of the international banking situation to the Manhattan Institute of Policy in October 1982. The address is a masterpiece of misdirection of attention. Instead of speaking of bankruptcy and defaults, he explains that the poor nations—called "developing," of course—are suffering from "liquidity problems." (So was David Copperfield's Mr. Micawber, who would have just loved the banker's address.)

The banker's most astonishing statement was the assertion that "There is a *negligible* risk of permanent default or debt denial in sovereign lending because sovereign borrowers cannot cease to exist." (The italics are the speaker's.) "Astonishing" is the mildest word one can apply to such a claim.

Investigation showed that the speaker was born in the year 1927, or ten years after the Russian revolution. He had no direct personal memory of that event—but surely he must have heard about it after he grew up?

Russian borrowers did not cease to exist, but their sovereignty certainly changed. The Bolshevik government repudiated *in toto* the debts of Czarist Russia. Human hope being what it is there was a brisk speculative market in Czarist bonds for another 20 years. (Most of the trades were handled by a New York broker named, curiously, Carl Marks.) In the end reality was too much for the speculators, and the beautifully printed bonds became just so much paper.

It is hard to believe that a man who has spent much of his life in banking would not have heard of the Russian default. Did our banker willfully choose to ignore this embarrassing history? Or was he genuinely ignorant of it? Cicero said that those who have no knowledge of what has gone before must forever remain children. It is a frightening thought that our bankers, who have at their disposal hundreds of billions of dollars, may be only children in their understanding of some aspects of human behavior.

Bankers, the New Commonists?

Bankers, like countless others who want to get ahead in the world, are sometimes tempted to compromise their principles. Faced with potentially monstrous losses in the international area they may hope to benefit from a new version of the time-tested CC–PP Game. If the foreign loans are paid off, they make profits in the usual way. But if poor countries default and our government steps in to make good on the payments, the banks collect their usual profits while the costs are imposed on the general public. The imposition takes place through several routes.

First of all, keeping interest rates high after 1973 meant that costs were imposed on every American who, directly or indirectly, borrowed money—which translates simply to every American. If and when an ad hoc F.D.I.C. for international loans is created, the costs of the defaults will be added to the already existing national deficit. This may result in higher taxes or (alternatively) in the "monetization" of the unpaid debts through deficit financing and subsequent inflation (the greatest of all hidden taxes).

Whatever euphemisms are used, costs will be commonized.

All this will be with the approval, if not the connivance, of a group of men ready at the drop of a hat to condemn "communism." Condemning communism, they are willing to practice commonism when need be. They need not be faulted for what they condemn, but they do deserve criticism for being willing to base their salvation on a basic plank of Marx's platform, namely making distribution in accordance with need.

CHAPTER 19

Renunciation Resurrected

By now it should be apparent that there is a kinship between the ecolate orientation and the conservative temper. The word "conservative" does not mean the same thing in different epochs; even within a single epoch, many meanings are in contention. Perhaps a conservative may be most simply defined as a person who doesn't want to change any traditional practice, no matter how patent its shortcomings.

In a world seldom challenged by scientific or technological novelties, this sort of conservatism may well be the highest wisdom. But that has not been our world for at least two centuries. It will not be our world for as far into the future as we can see. Even if the pace of scientific discoveries should slow to zero, the recent past has left us with such a mass of undigested technologies—material technologies for which we have not yet found matching political techniques—that trying to stand pat won't do.

The conservatism that is *chic* among the movers and shakers of our time can be called economic conservatism, which is designed to preserve the social arrangements that made it possible for the present rich to become rich. For those who played a strict free enterprise game a good defense can be mounted. But many became rich by playing the CC–PP Game. It is not easy to find a defense for their activities that is not also a defense of those who benefit from the welfare state they abhor. So the Double C–Double P wealthy use language that disguises the rules by which they play: typically, they say they are defending and conserving private enterprise. A medieval highwayman could have claimed as much.

Conservatism as Humility

Ecology stands in opposition to any economic conservatism that tolerates the CC–PP Game. Yet one of our most perceptive political scientists, William Ophuls, has identified ecology as "a profoundly conservative doctrine in its implications." Such a characterization must come as a shock to those who usually consider ecologists to be flaming radicals.

The justification for labeling ecology "conservative" can be made clear by examining the defenses for older forms of conservatism. No one has epitomized political conservatism better than Edmund Burke, whose speech on "Conciliation with the Colonies" justified his activity in preventing political change by a eulogy of the British system of government as the product of

> the ancient, rustic, manly, homebred sense of this country. I did not dare to rub off a particle of the venerable rust that rather adorns and preserves, than destroys, the metal. It would be a profanation to touch with a tool the stones which construct the sacred altar of peace. I would not violate with modern polish the ingenuous and noble roughness of these truly constitutional materials. Above all things, I was resolved not to be guilty of tampering, the odious vice of restless and unstable minds. I put my foot in the tracks of our forefathers, where I can neither wander nor stumble.

Thus did Burke identify a major meaning of the word "conservative," though the label was not so applied until 1830, thirty-three years after Burke's death.

Burke's justification of the conservative position does not appeal to many moderns. To speak of the "venerable rust," the "sacred altar," and the "ingenuous and noble roughness" of social institutions now loses, rather than gains, the support of the average man or woman. In our time it is the new, not the old, that easily commands unquestioning acceptance.

Tradition needs a better defense than its antiquity. The best defense, according to the economist F. A. Hayek, is this: tradition almost invariably works better than the latest social or economic innovation. Hayek goes on to say: "One of the most important tasks of our intelligence [is] to discover the significance of rules we never deliberately made . . . the obedience to which builds more complex orders than we can understand."

What is called "liberalism" in our day is imbued with the "Can do!" spirit, the confidence that we can alter the traditional arrangements of things to create a new world that is "nearer to the heart's desire." If a conventionally condemnatory moral term is to be attached to liberalism it must be something like *superbia—hubris*—pride. By contrast, conservatism of the sort that suffuses the writings of Burke and Hayek merits the name *humilitas*. Until modern times, humility has been considered a virtue.

The ecolate view of the world springs from a deep humility. The First Law of Ecology, "We can never do merely one thing," warns us that any human intervention in the order of things will likely have unforeseen consequences; and that many, perhaps most—perhaps all—will be contrary to our expectations and desires. The warning of the First Law is of a piece with Francis Bacon's warning that "Nature, to be commanded, must be obeyed." The spirit of this advice contrasts sharply with the attitude of our movers and shakers, who presumptuously label their *interventions* "development," "reclamation," or "improvement."

The word "development" applied to man-made alterations of the environment presumes a preexisting program that in fact does not exist. The word "reclamation" applied to the destruction of an estuary presumes that human beings once controlled this natural resource and that we are now merely reclaiming what has always rightfully been ours—a blatant lie. "Improvement," as the word is used by real estate operators, presumes that anything that is taxable is preferable to something that is not. Language in the service of Ego.

Ecolate conservatism, like political and economic conservatism before it, seeks to conserve ancient goods; but the goods of ecology are immensely more ancient. Few political traditions are more than a century or two old. Some economic traditions are older. (But do we really want to preserve the highwayman's tradition of commonizing costs and privatizing profits?) The goods ecologists and environmentalists want to conserve come to us from millions of years ago: the beauty and wisdom of self-balancing nature.

Is the human spirit uplifted by the razing of a redwood grove? Are autumn days made more beautiful by filling in and destroying wetlands that once attracted millions of migratory birds? Is the world made lovelier by loading our ship of state to the scuppers with human beings if the populations of tomorrow will thus be forever deprived of contact with hundreds of thousands of now existing species of plants and animals?

The Confidence of Science and Technology

The humility of classical political and economic conservatism does not sit well with technologists who are, in Burke's quaint words, committed to the "odious vice" of "tampering" with nature.

The nineteenth century brought fantastic progress in the devising of new machines to perform old tasks. Problems that were long thought insoluble were solved. In the area of technology humility was replaced by arrogance as one astonishing invention succeeded another. Whitehead has said that the greatest discovery of the nineteenth century was the invention of the method of invention. Once this process had gained momentum those who called for caution in the adoption of new technologies were scornfully told, "You can't stop Progress!"

Were the Luddites Wrong?

In 1811, in the region of Robin Hood's Nottingham, there was a revolt against the introduction of labor-saving textile machinery. Displaced workers, marching under the banner of a mythical King Ludd, smashed the machinery. Soon the state, at the behest of the wealthy classes, smashed the smashers. In five years the Luddites, as they were called, were routed. Machinery came in and workers moved out, some to other lawful occupations, some to crime, some to starvation.

Conventional wisdom has it that the Luddites were wholly wrong, because they opposed a technological imperative. The use of more efficient machines was defined as Progress, no matter how much some people might suffer. In deciding where the balance of truth lies we should keep in mind that history is written by the winners. Are we sure the winners were right? The total effects of newly introduced machinery included loss of jobs, conversion of geographically fixed labor to migratory labor, dissolution of families, and the massing of workers in central factories instead of dispersed in separate cottage industries. If a comprehensive accounting of the effects of the new machinery were carried out, would it justify the change?

No one knows. The ecolate way of looking at a total system came long after the Luddites. No critical searchlight was focused on "side effects" in the nineteenth century. It was some time before it was realized that "all effects are effects, period." Academic learning was the product and property of classes that stood to benefit by every labor-saving invention, no matter what its total effects. In the absence of welfare legislation commercial enterprisers managed to privatize the profits of innovation among their own kind while commonizing the costs of unemployment and poor health among the lower classes.

Though conventional wisdom now holds that all resistance to the introduction of new inventions is Luddism, and that Luddites are always wrong, I don't think it can be maintained that we yet know the final answer. It cannot yet be confidently asserted that industrial societies operating under the rules of free enterprise have solved the unemployment problem. It is a brutal truth of our time that only war plus free enterprise has produced full employment. But we can no longer afford a major war. The large amount of hidden unemployment in the form of socially wasteful work, and the social disorder that is in part attributable to the excessive mobility of labor and families, merit a thorough and objective study before making a final judgment on Luddism. Such a study will probably never be made. We will just have to muddle along without it.

A New Fatalism?

In the past we in the European tradition have made fun of "primitive" societies that rejected change. We have accused their leaders of fatalism when they assumed that any change would necessarily be for the worse. But what of our slogan "You can't stop Progress"? Is this not a fatalistic statement? It postulates a Juggernaut that men and women are powerless to stop, though the engine is of our own making.

Our admirable "Can do!" attitude in the face of technological problems has, curiously, no equivalent in the political realm. Unbounded optimism in the first area is almost universally coupled with complete pessimism in the second. We see an example of this coupling in the word "shortage," which neatly combines technological optimism with total political pessimism. The explanation of this curious state of affairs is worth looking into.

"Longage," a Tabooed Word

"Philosophy," said Ludwig Wittgenstein, "is a battle against the bewitchment of our intelligence by means of language." Though Wittgenstein was a twentieth-century figure, there is no reason why we cannot, in this connection, think of "philosophy" in its ancient meaning as the love of learning—all sorts of learning. A mind bewitched by words has difficulty thinking about the reality for which words stand as deputies. The bewitching word "shortage" has imposed tunnel vision on many an intelligence.

There are approximately five billion people in the world today, and a billion of them would feel more comfortable if they had more food. After realistically allowing for some waste and imperfect distribution we are forced to conclude that the amount of food is not quite in balance with human demands. We summarize the situation by saying: "There is a shortage of food."

Why don't we say, "There is a longage of people"? For one thing, the word "longage" does not exist (as far as dictionaries are concerned). The word "shortage" seems to have been coined in the Chicago grain market around 1868. More than a hundred years have passed since then and still people do not speak of "longages." Why not?

The question brings us back to the principle that language is action. Our choice of words betrays the kind of action we are willing to take. The word "shortage" is welcomed by everyone who stands willing to supply more of something—at a profit. More wheat, more transportation, more distribution systems—there is a profit to be made in supplying every single "more." But who can make a profit from supplying less? Who benefits by teaching people to use less? Who profits from selling temperance? Why speak of "longage" if you can't make money out of it?

As concerns population I think there is another reason why we automatically suppress the thought of a longage of population: we fear the action that might be proposed. At a subconscious level we picture some heartless dictator lining up the surplus people and shooting them. Such a remedy is clearly unacceptable.

In fact, however, population can be reduced by an equally effective measure that is not violent: attrition. Let the birth rate be reduced, by nonviolent means, to a level below the natural death rate. In time, population will be reduced just as effectively in this way as it would be by violent action. Properly managed, attrition should be acceptable. (Morally acceptable, that is, though politically difficult to bring about.)

The idea of attrition may be too subtle for the subconscious mind to deal with. Freud called our attention to the fact that the subconscious mind cannot deal with negatives, and attrition has the attributes of negativism. Thus it comes about that the potentially useful word "longage" is throttled at birth. To live a thoroughly rational life we must transcend the limitations of the subconscious mind.

What Renunciation Means

The clamorous "Can do!" spirit that grew out of engineering in the nineteenth century was, paradoxically, accompanied by the rise of a "No can do" spirit in the exact sciences, on which engineering depends. Only "Can do!" captured the imagination of the general public. Appreciation of the importance of "No can do" was confined largely to professional scientists; even among these, full understanding came only with the twentieth century.

The sorts of things the "No can do" spirit deals with are the laws of thermodynamics and all scientific conservation principles. There aren't many of these, but they have profound implications for the rest of science. Because technological discoveries have surprised us in the past, people outside science are inclined to suppose that every scientific law will sooner or later be overthrown. Scientists demur. They perceive a hierarchy of laws, with those at the base of the intellectual structure being almost certain to survive all revolutions. The reasons for this confidence are subtle.

In 1942 the mathematician E. T. Whittaker called the basic "No can do" assertions *postulates of impotence*, saying that each such statement "asserts the impossibility of achieving something, even though there may be an infinite number of ways of trying to achieve it. A postulate of impotence is not the direct result of an experiment, or of any finite number of experiments; . . . it is the assertion of a conviction of the mind, that all attempts to do a certain thing, however made, are bound to fail."

True perpetual motion is impossible. Matter/energy can be neither created nor destroyed (and there is a fixed equivalence of matter and energy). . . . Scientists agree with these statements, but some are made uneasy by the term "impotence." The physicist R. B. Lindsay proposed that we speak instead of "postulates of renunciation." There is something to be said for both alternatives.

Scientists have renounced attempts to create a perpetual motion machine, to create energy, and to square the circle or trisect an angle with only a straight-edge and a compass. To some outsiders such renunciations make scientists seem intolerant and narrow-minded. The fact is that any scientist or mathematician who initially gives attention to circle-squarers and the like sooner or later is driven to close the door of his office, saying "Enough is enough!" The question, "Why don't you keep an open mind?" is met with the counterquestion, "What's eating you that you should recommend that I butt my head against a stone wall?"

Impotence postulates are cast in negative form but they are not sterile. The working out of their implications has led to great practical developments. Engineers, despite their general loyalty to the "Can do!" spirit, generally refuse even to look at blueprints for a perpetual motion machine. This intolerant attitude has made the lives of engineers productive.

The scientist who refuses to look at plans for a perpetual motion machine is like the investor who refuses to put money into a business where the books don't balance. To avoid the discipline of balancing the accounts, whether in business or science, is to ensure that chaos will take over. Technological man renounces some dreams that he may achieve others. Only if some things are impossible can other things be. This is the deep faith of sane science.

Science, the most progressive of all human activities, is founded on renunciation; perhaps it is time that we extend this approach to other areas. Perhaps some of the difficulties that now seem insuperable can be overcome if we identify the desires we can afford to renounce.

A civilization is defined by the dreams it renounces. Whenever there is a mismatch of supply and demand there are two ways of dealing with the situation in words: we can speak of either a shortage of supply or a longage of demand. The alternatives are, in logic, equally valid. But for a long time our civilization has repressed the thought of longages of demand. The suppression may have been for good practical reasons, but all acts of suppression end up by stunting the mind. A basic insight of ecology is that there are real limits to the world (though we sometimes make mistakes in defining them). The recognition of limits makes possible the development of accurate methods of accounting, without which no honest description of the world is possible. Historically, the belief in infinite supplies has greatly stimulated innovation. As we approach the limits this belief becomes less fruitful. Finally it becomes dangerous. When our civilization renounces the dream of infinite supplies it will move into a new and saner mode.

The new mode of civilization will be closer to that of the traditional societies we used to label "primitive." The burden of proof will be placed on those who propose change, not on those who resist it. Already we are moving in that direction with our laws governing the introduction of new medical drugs and of "developments" that alter the common environment.

The new mode will be significantly different from the earlier traditional forms of conservatism. The new conservatism will distinguish between conserving the material base of society and conserving information, which is nonmaterial. Because the scientific laws of conservation apply only to matter and energy, in the crowded world of tomorrow more attention will have to be given to ensuring that an innovation does not threaten to worsen the available supply of these two entities. But since information is not conserved, in the scientific sense, we can look forward to a society that has become more adventurous in devising new ways of shaping the management of the material world to the exigencies of human nature. Among other things, this will mean that the word "longage" will have to become an operative term in the public vocabulary.

CHAPTER 20

By Way of Summary

In developing the implications of ecology for human affairs I have tried to follow the example set by Spinoza. His words bear repeating: *I have labored carefully not to mock, lament, or execrate human actions, but to understand them; and to this end I have looked upon passions, such as love, hatred, anger, envy, ambition, pity, and other perturbations of the mind, not in the light of vices of human nature, but as properties. . . .*

Not vices, but properties—these are the aspects of human behavior that repay the closest study. But so pale an abstraction as "properties" may not be enough to call forth our best efforts; if that is so—if we need a more solid antagonist—I suggest that Folly will serve better than Vice.

As concerns their final effects, the reactions of a self-sustaining system are often ambivalent. Quickness to anger can save a man's life in one situation but lose it in another. The immune reactions of the human body may cure a disease or produce a disabling allergy. A jealous defense of "national honor" may enhance a nation's standing in the world, or set in train a sequence of knee-jerk responses that end in destroying the body politic.

The ambivalence of reactions leads to speaker-biased pairs of adjectives. "I am firm, you are bull-headed. You are fickle, I am adaptable. I am prudent, you are stingy." The Siamese-twinning of vice and virtue is repeated again and again in traditional language. Once a person becomes sensitized to the perception-bending nature of labels, their emotional power is lessened. Note that Spinoza identified both love and pity as "pertubations of the mind," thus implying that we should question the value of even these highly prized abstractions.

When we elect to judge acts by their consequences we can no longer blindly accept love and pity as pure virtues.

The pursuit of science has two objects: understanding and control. To achieve these ends we must master more than words: we must also know quantities and have some comprehension of the total environment of forces. Literacy must be augmented with numeracy and ecolacy. Beyond purely literate "vice" and "virtue" lies reality, which is inescapably both numerate and ecolate.

In the reworking of education it is time to supplement the Three R's with the Three Filters. The skills of Readin', Writin', and 'Rithmetic need to be combined with an attitudinal checklist that asks if the best words have been used, if quantities have been duly considered, and if the consequences of time and repetition have been taken into account. The "bottom line" of an analysis needs to be subjected to filtration that is simultaneously literate, numerate, and ecolate. No single filter is sufficient for reaching a reliable decision, so invidious comparisons between the three is not called for. The well-educated person uses all of them.

In spite of its name, numeracy is concerned with more than numbers. The *relative* size of quantifiable factors is often more important than their exact measures. The importance of scale effects can be appreciated with little actual measurement. The existence of quantitative limits must, in general, be granted even though it is often not possible to state them with numerical exactitude.

But numbers are not enough. Oscar Wilde defined a cynic as "a man who knows the price of everything and the value of nothing." This definition would do equally well for the most narrowly numerate person. Value is a relative concept: the value of each action is determined by comparing it with other possible actions. Every measured thing is part of a web of variables more richly interconnected than we know. We use the ecolate filter to ferret out at least the major interconnections. Every proposal of a plausible policy must be followed by the question "And then what?" Not until we have asked this question (and answered it to the best of our ability) are we ready to put a plan into action. The ecolate question being open-ended, we can never completely answer it. When action is inescapable we must keep in mind the thought that our analysis of the situation may yet be mistaken.

Finally, there is the filter that is historically the oldest, the literate filter, where that term is understood to include more than reading and writing. Words, whether vocalized or printed, are indispensable in explaining the insights of numeracy and ecolacy to others. In the past many wordsmiths have been neglectful of the numerate and ecolate aspects of things. In the science-saturated world of the future they should recognize that melding numeracy, ecolacy, and literacy is both their civic responsibility and their professional opportunity.

The talent for handling words is called eloquence. Talent is always desirable, but the talented may have an unfair, even dangerous, advantage over those with less talent. More than a century ago Ralph Waldo Emerson said, "The curse of this country is eloquent men." The curse can be minimized by using words themselves to point out the danger of words. One of their functions is to act as inhibitors of thought. People need to be made allergic to such thought-stoppers as *infinity*, *sacred*, and *absolute*. The real world is a world of quantified entities: "infinity" and its like are not words for quantities but utterances used to divert attention from quantities and limits.

That education needs to put more emphasis on mathematics and science is now widely recognized. Whatever our various opinions as to the ultimate feasibility of a unified world-state, most of us recognize that, as far down the road as we can see, the world will continue to be an assemblage of competing nations. We hope that the struggle can be kept below the level of military action, but self-interest operating in a world of limits dictates that we continue to adjust to the demands of competition, whether its form be commercial, scientific, or something else.

Americans should give close attention to the structure of education in other nations: their young are our next competitors. We cannot afford to overlook the fact that in Japan, for instance, a recent survey showed that some 2,000,000 high school students are studying calculus, as compared with only 165,000 in American high schools. This, in spite of the fact that America's population is twice the size of Japan's. In proportion to their population, the Japanese studying calculus in high school are 23 times as numerous as Americans. Such a difference cannot be trivial. Japan is only one of many countries with which we will have to compete in the future.

Some will object that mathematics is only one specialty among many, so we shouldn't make much of this difference. I do not agree. Mathematics is at the heart of science, and science is indispensable to the survival of a great nation. (Tiny nations can survive by borrowing the fruits of science from others.) Properly taught, mathematics and science deserve to be placed among the humanities. It must be admitted, however, that mathematics and science are often not properly taught. Riding herd on education is as difficult as any other communal task. Achieving quality takes money; maintaining it takes vigilance.

Since the Renaissance intellectual leaders have struggled to reestablish our civilization on a humanistic foundation. "Humanism" has many definitions, and there is no universally recognized authority to say which is right. I think four words of Alexander Pope give the essence in a nutshell: *To err is human.* Nothing is so contrary to the spirit of humanism as a belief in the existence of absolute reliability. Moralists demand it, promoters promise it—but scientists study its rhetorical opposite, *un*reliability, under such headings as statistics, design of experiments, and decision theory. Humanism aims to understand the all-too-human sources of unreliability and to learn how to survive and enjoy life in a complex world shot through and through with error.

We must learn to live happily with less than we can dream of. The true humanist can have no more than a qualified enthusiasm for Emerson's advice to "Hitch your wagon to a star." Fine advice, if trying to do the impossible evokes the best there is in you. But breaking your neck is also a possibility.

Going from personal to community ideals, it is essential that the body politic reject the underlying philosophy of the Delaney Amendment, which mandates that the amount of a carcinogenic contaminant of foodstuff shall be zero. Many environmentalists are attracted to the Zero Tolerance philosophy, wanting to bring into being an absolutely pure atmosphere and water supplies absolutely free of all pollutants. The greed of some enterprisers in seeking profits through pollution is matched by a different sort of greed of some environmentalists in demanding absolute purity regardless of cost. Both aberrant groups need to be reminded that greed was long ago rightly judged to be one of the Seven Deadly Sins.

We need to understand why hitching the environmental wagon to the star of Zero Tolerance is unwise. If the cost of purifying—removing a pollutant or contaminant—is graphed against the degree of purity of a material, it is found that the curve finally turns sharply upward. For a single example, consider the cost of alcohol. In its usual commercial form it is "contaminated" with 5 percent water. "Absolute alcohol" (supposedly 100 percent pure) is much more expensive. Removing the last 5 percent of the contaminant increases the cost by 117 percent. (These calculations are based on the 1984 costs of laboratory alcohol. The cost of legally potable alcohol includes a high tax, which has no connection with the cost of production.) Evidently the cost-purity curve takes a sharp turn upward as we approach absolute purity. This is a general principle applying to all substances.

The more sensitive the chemical analysis, the smaller the fractions of contaminants that can be discovered. To remove increasingly minute fractions costs more and more; but the benefits of removing them are less and less. We must recognize that the world is finite; the resources available per unit time are finite. When costs are paid out of a common pot, extreme purity in one dimension can be achieved only by impoverishment or contamination of others. Trying for too much we achieve less. Rational limits must be set to every ideal of purity.

We need to acknowledge explicitly that we are always going to have to live with pollution. We must convert each purely literate standard to a numerate one so that it can be subjected to reexamination whenever new methods and new cost-purity curves are developed. When all the facts are in and all the evaluating has been done, the setting of each limit requires the drawing of an arbitrary line for the purity to be demanded. We must learn to accept the arbitrary.

Temperance is required even of the passion for justice. Perfect justice is as impossible as absolute purity. Individual differences in the reaction to competitive challenges, coupled with the positive feedback of power—"To them that hath shall be given"—ensure that inequities of distribution will reappear no matter how often they are ironed out of the politico-economic system. It is dangerous to think of poverty only as a state or condition; poverty is also a process. Poverty-minimizing policies are most successful when they deal with the process rather than the state. We must recognize that all control operations incur costs; excessive controls generate their own kind of poverty. We must face two questions "How much are we willing to pay to achieve a specified amount of good?" And "Who is the 'we' that is called upon to pay?"

The classic question, "Who benefits? Who pays?" splits the "we" into two "whos." It raises problems in the distribution of income and wealth. The traditional Christian-Marxist ideal, "From each according to his ability, to each according to his needs," implies a common pot to be dipped into at will. Within a loving family this ideal may be the best, but it is a singular ethics that blinds us to the fact that such singulars as "his" cannot safely be replaced with plural subjects. "The family of man" is a dangerous myth.

At the scale of a small, intimate community a Christian-Marxist distribution system may work well enough, but (as engineers would put it) this distribution system "doesn't scale up well." Shame, the necessary psychological enforcer of a commonistic system, is effective only in small communities. When a community is "large," say 150 or more, shame buckles under the ubiquitous pressure of egotism. Asserted needs escalate, apparent abilities wither. By the very logic of the commons, individuals are paid to do wrong: operational responsibility becomes negative. Since shame cannot control asserted needs when the numbers are great, the community has only two paths open to it. It can allow individual freedom, in which case the escalation of asserted needs brings ruin to all: this is the tragedy of the commons. Or the state can define the needs of every individual: this is tyranny.

Commonistic systems are ubiquitous but seldom labeled for what they are. Correctly identified they would soon be rejected. The thought-inhibiting function of language is employed to get commonistic systems accepted under other and approbative names. Two different languages are employed by those who support commonistic arrangements: the language of compassion and the language of economics. Those who want to better the plight of others use compassionate language. Those who want to benefit themselves or their firms use—or rather, misuse—the economic language of free enterprise. The two linguistic strategies deserve to be exposed.

Compassionate attempts to shield individuals against the "slings and arrows of outrageous fortune" have produced a multitude of insurance schemes, both private and public, both voluntary and compulsory. A 100 percent supplying of "needs" from the commons produces ruin at the end of a runaway process. Because of arsonists, ruin is a perpetual threat to fire insurers. It is also a threat to nations with too generous a view of the medical needs of their citizens. One protective response is to pay less than 100 percent of the cost of satisfying needs; such a policy gives the insured party a piece of the action and hence a reason to try to contain costs.

The international sharing of wealth that goes under the name of Foreign Aid obeys the same rules. In the short run such sharing creates a Zero-Sum Game: one country's loss is another's gain. Wealth is shared. In the long run, however, gifts inhibit private initiative and encourage further population growth in the recipient nation. In the long run, satisfying needs through commonization increases "needs." Thus a Negative-Sum Game develops as per capita wealth is lowered in both donor and recipient countries. What starts as a sharing of wealth ends up as a sharing of poverty.

A striking visualization of the international problem was given us by a photograph made by NASA, the space agency; but publicists seized on the wrong photo. The too often printed view of the entire earth as seen from space led to some lovely rhetoric about "this small and blue and beautiful sphere with its swirling clouds, floating in a sea of darkness, surrounded by eternal silence, mute evidence that we all are, and of necessity must be, brothers." (I meld several similar statements together.) Because the space program had given us the ability to capture the entire image of a planet with a diameter of 7,900 miles on a single photo, taken from a sufficient distance, it was now asserted that the earth's wealth must be commonized and shared by all. Rhetoric, not logic, connects the observation with the conclusion.

The NASA photo that should have engaged our attention was one taken from near-space that showed a green pentagon in the midst of the brown Sahel of Africa. Because of the low level from which this photograph was taken, aberrations caused by the earth's atmosphere led to a picture that was not as sharp and beautiful as the blue globe photographed from distant space, so the Sahel photo was less widely printed. It was, however, incomparably more significant in its implications.

Inside the green pentagon was grass enough for the animals pastured on it; outside was ruin and devastation for both man and beast. Why? Investigation at ground level showed that the pentagon was enclosed by a fence. The inside was owned and controlled to ensure that the number of animals never exceeded the proper carrying capacity of the environment. Outside the fence the Christian-Marxist rule, "to each according to his needs," prevailed, and the uncontrolled and irresponsible escalation of animal and human populations ruined the common heritage. The Sahelian picture is a symbol of what will happen to the entire world if humanity decides that needs create rights. This is the picture from space from which morals should have been derived.

Keeping our feet firmly planted on the earth we should note the economic strategies used in introducing commonistic systems into human affairs. The word "subsidy" is the nearest to frankness that enterprising promoters come. A subsidy is a means of dipping into the commons of the public treasury without acknowledging commonistic intent. The subsidized enterpriser seeks to establish a game in which costs are commonized while profits are privatized—the classic and unacknowledged CC–PP Game.

The number of different Double C–Double P Games that have been invented is beyond counting. Lumber firms manage to get the U.S. Forest Service to pay most of the cost of raising trees to harvestable size. Cattle ranchers are allowed to pasture their animals almost cost-free on public lands. Tobacco farmers are paid to produce a product that diminishes public wealth through the resultant medical bills for lung cancer. Public utilities are subsidized in a multitude of ways to produce electricity through nuclear disintegrations—and then the costs of taking care of the wastes are paid for out of the public till. A federal "superfund," financed in part out of the public treasury, is used to pay for cleaning up landfills contaminated over decades by private firms "throwing away" their wastes. Domestic losses from failed banks are absorbed by the Federal Deposit Insurance Corporation, which can dip into the public treasury. And private banks show little prudence in making massive loans to poor foreign countries since the bank managers are confident that these losses too will be taken care of by the nation as a whole.

The enterprises described above—and they are only a fraction of a long roster that could be assembled—are all examples of the CC–PP Game, of commonizing the costs while privately sequestering the profits. Yet they are all presented as examples of "private enterprise at work." (So they are: private enterprise at work on the public pocketbook.)

We have moved a long way from the nineteenth century, when people supposed the whole meaning of commercial success was to be found in a saying attributed to Emerson: "If a man makes a better mouse-trap than his neighbor, though he build his house in the woods the world will beat a path to his door." Nowadays the art of commercial success too often seems to revolve around techniques for mouse-trapping the government.

Though the language used by enterprising cost-commonizers is derived from economics it should not be supposed that this usage has the blessing of academic economists. It is perfectly obvious that commonization is not a form of free enterprise, as many economists have pointed out in their public statements.

The gap between academia and the public often makes possible the serious misuse of academic language. Academic biologists, for example, are disturbed by the claim that anti-evolutionists are developing a new "Creation Science," when it can be shown, by their own words, that they are engaged in creating an antiscience (a conclusion reached also by Judge Overton in the 1982 Arkansas decision). The claim of enterprising cost-commonizers that they are engaging in free enterprise is equally illegitimate. Both biologists and economists have a massive job of public education ahead of them.

The evils of commonization are many, but this should not blind us to the fact that some forms of commonization are desirable. The cost of laying down and maintaining public sidewalks is commonized, yet no one complains. (The management costs of billing each individual for his use of this commons would make a truly responsible system more expensive for everyone.) The situation of motor roads is marginal: sometimes we set them up as toll roads.

Scientific research presents an interesting problem. Ideas, by their nature, become part of a commons of information. Commonized gains don't motivate individuals very well. To encourage people to try to work up good ideas we create forms of private property called patents and copyrights. This economic motivation works for innovations that can be quickly exploited. Research that is labeled "R&D"—Research and Development—falls into this category.

But the purest of "pure research" is unlikely to yield commercial returns in this generation. Expected economic benefit to be derived in the future must be discounted to the present by the standard method of economics. The "present value" of the postulated future gain must be compared with the very real present cost of the research. Even a modest interest rate used in calculating "present value" results in discounting a distant future good nearly to zero. Comparing that with a sizable present cost of research leads the rational "economic man" to conclude that no one can afford to subsidize pure research.

Yet there are counterarguments. In general it is true that nations that have the most vigorous programs of pure research also have the most productive R&D. The financial support of pure research creates an intellectual climate that favors all kinds of research. Nations that support pure research get the jump on others. A nation that adopted the most rigid economic attitude toward pure research, financing none of it, would almost certainly find itself behind the eight ball a generation later. So, "economic" or not, pure research needs supporting, for the sake of the future.

The ordinary private enterpriser cannot afford to support pure research: he has to follow the simplest economic rules. An enterprise that is so large as to be above much of the competition may be able to indulge in the luxury: General Electric and AT&T supported some very fine pure research for most of the twentieth century.

Most pure research, however, has been supported in two ways. First, by drawing on the commons, through state-supported universities and colleges, through government grants (e.g., grants made by the National Science Foundation), and through government agencies (though the record of the latter is not impressive). Second, through private philanthropy. Private universities and private foundations pay for pure research out of funds that were earned by private enterprisers in an earlier day and set up as a public trust for doing things that tax-supported agencies find difficult to justify. Though the managers of the granting institutions can ignore "economic man" calculations, they nonetheless try to estimate, in some intuitive way, the probable future gain flowing from proposed research. The future that is plugged into these rough, intuitive equations is often beyond the expected lifetime of the decision maker. The ability and willingness to make such decisions oriented to the distant future is a remarkable characteristic of the human animal, one of which we can rightly be proud.

A strong future orientation is an essential feature of what we call "conservatism." This much abused word has had many definitions, changing with time and circumstances. None can be said to be uniquely correct, but something useful can be said about the more often used definitions.

To be a conservative is to want to conserve, preserve, keep, save or maintain—*something*. But what? In the days of the landed aristocracy, in Edmund Burke's century, the conservative wanted to keep intact the handsome country estates, which meant guarding the special privileges of those who were the stewards and beneficiaries of these great domains.

As the rising commercial class displaced the landed aristocracy in power, another form of conservatism found its voice in Adam Smith. The production of national wealth depended on a multitude of economic transactions taking place in a complex web of social relationships. Attempts by the Mercantilists to organize this web rationally and explicitly under the direction of the state produced systemic malfunctions time after time. This led to the suspicion that leaving the operators alone might be better than national planning: thus was born the doctrine of *laissez faire*. Smith gave this doctrine its most memorable justification:

> As every individual, therefore, endeavours as much as he can both to employ his capital in the support of domestic industry, and so to direct that industry that its produce may be of the greatest value; every individual necessarily labours to render the annual revenue of the society as great as he can. He generally, indeed, neither intends to promote the public interest, nor knows how much he is promoting it. By preferring the support of domestic to that of foreign industry, he intends only his own security; and by directing that industry in such a manner as its produce may be of the greatest value, he intends only his own gain, and he is in this, as in many other cases, led by an invisible hand to promote an end which was no part of his intention. Nor is it always the worse for the society that it was no part of it. By pursuing his own interest he frequently promotes that of the society more effectually than when he really intends to promote it.

Note that Smith claims no more than that the enterpriser "frequently" works for the public good as he seeks to maximize his own returns. Smith did not claim an absolute good for freedom in any and every kind of enterprise, but his arguments did open people's eyes to the melioristic tendencies of many self-serving economic activities. Ambition among primary producers (e.g., farmers) results in the production of more of their product, and hence lower prices to consumers. Price wars among middlemen, though they may bankrupt individual merchants, benefit consumers by keeping prices down. Even the hectic activities on the stock exchange, though no substantive product is produced, benefit the whole economic system by speeding up the adjustments of prices to new information. Such auctions act like lubricants in the economic system, for which the lubricating agents are paid their fee. Fair enough, it seemed. Smithian explanations help to diminish the force of envy, always a threat to domestic peace. Anyone who preserves the peace deserves, in some sense, to be called a conservative.

Under laissez-faire rules, economics "frequently" produces

more prosperity than the most carefully planned systems men can think of. But "frequently" is not "always," and compassionate critics may yearn for reform. But reform action taken in advance of deep understanding is perilous. The Nobel laureate economist F. A. Hayek has repeatedly sung the praises of the adaptedness of the unplanned economic order. In his words: "It may still be one of the most important tasks of our intelligence to discover the significance of rules we never deliberately made, and the obedience to which builds more complex orders than we can understand."

Yet it does not follow that all rules that grow like Topsy necessarily make things better for everyone. For someone, yes; otherwise the rule would not grow and prosper. But to identify as "conservative" the support of all rules that have for a long time enriched limited classes of people is to assume that which needs to be proved.

Because private enrichment follows from pursuing the Double C–Double P Game, it does not follow that commonizing the costs deserves the name of conservatism. Commonism, being negatively responsible, encourages the waste of resources and brings about a tragic ruin for all. The individual who profits by playing the CC–PP Game is not the individual praised by Smith as one who "necessarily labours to render the annual revenue as great as he can." The commons game, with its built-in negative responsibility, pays the individual (in the short run) to reduce the annual revenue for everyone (in the long run).

In the third century B.C., Zeno of Citium began his career as a philosopher by assuming that human happiness can be best achieved by living "in agreement" with oneself. Thus did this Stoic give new expression to the motto "Know Thyself" inscribed earlier on the temple of Apollo at Delphi. As he grew older and wiser Zeno decided that it was also necessary "to live in agreement with nature." A Stoic rewriting of the Delphic motto could well be: "Know Thyself, and Thy Place in the Natural World."

The intellectual development of the Western world in the

scientific era, particularly in the past century, can be seen as a repetition of Zeno's personal history. The world-view that was first built on an exclusively individualistic basis is now seen headed for disaster unless we take account of nature, where the word "nature" refers not only to the external material world of plants, animals, rocks, and all that, but also to the nonmaterial world of logical relationships. "We get what we pay for"—so we had better make sure that we pay for the right behavior.

The political scientist William Ophuls has said, "Ecology is a profoundly conservative doctrine in its social implications." The garden-variety conservative of our day is all too often a conservative only in the sense that he wants to preserve every societal arrangement that has in the recent past enriched some enterprisers (including himself), without asking about the total consequences of the rules by which the profits were made. An ecoconservative, in contrast, out of a profound concern for the survival of the real wealth of the biological world and a passion for taking time seriously, rejects any arrangement that betters the individual by sacrificing the interests of all others, particularly the interests of posterity. To state the obvious, in the long run, the society that survives is the society that develops and conserves rules that serve the long-term interests of society.

A coldly rational individualist can deny that he has any obligation to make sacrifices for the future. By contrast, those who, for whatever reason, regard the resources at their disposal as an inheritance from the past that they feel obliged to pass on to their descendants, have a better chance of producing future generations prosperous enough to be able to continue to wrestle with the problems of increasing the quality of life.

Continuity is at the heart of conservatism; ecology serves that heart.

INDEX